Operations

July 2019

United States Government
US Army

Contents

		Page
	PREFACE	iii
	INTRODUCTION	v
Chapter 1	**MILITARY OPERATIONS**	**1-1**
	An Operational Environment	1-1
	War as a Human Endeavor	1-4
	Army Strategic Roles	1-5
	Unified Action	1-6
	Land Operations	1-9
	Seize, Retain, and Exploit the Operational Initiative	1-11
	Readiness Through Training	1-12
Chapter 2	**OPERATIONAL ART**	**2-1**
	The Application of Operational Art	2-1
	Defeat and Stability Mechanisms	2-4
	The Elements of Operational Art	2-5
Chapter 3	**THE ARMY'S OPERATIONAL CONCEPT**	**3-1**
	Unified Land Operations	3-1
	Decisive Action	3-1
	Consolidate Gains	3-5
	Activities to Consolidate Gains	3-6
	Principles of Unified Land Operations	3-7
	Tenets of Unified Land Operations	3-12
	Successful Execution of Unified Land Operations	3-13
Chapter 4	**OPERATIONS STRUCTURE**	**4-1**
	Construct for Operations Structure	4-1
	Operations Process	4-1
	Combat Power	4-2
	Army Operational Framework	4-2
Chapter 5	**COMBAT POWER**	**5-1**
	The Elements of Combat Power	5-1
	The Six Warfighting Functions	5-2
	Organizing Combat Power	5-7
Appendix A	**COMMAND AND SUPPORT RELATIONSHIPS**	**A-1**
	GLOSSARY	Glossary-1

DISTRIBUTION RESTRICTION: Approved for public release. Distribution is unlimited.

*This publication supersedes ADP 3-0, dated 6 October 2017, and ADRP 3-0, dated 6 October 2017.

Contents

REFERENCES **References-1**
INDEX **Index-1**

Figures

Introductory figure. ADP 3-0 unified logic chart vi
Figure 2-1. Army design methodology 2-3
Figure 2-2. Operational approach 2-4
Figure 3-1. Decisive action 3-3
Figure 5-1. The elements of combat power 5-1
Figure 5-2. Command and control warfighting function 5-3
Figure A-1. Chain of command branches A-2
Figure A-2. Joint task force organization options A-4
Figure A-3. Example of a joint task force showing an Army corps as joint force land component commander with ARFOR responsibilities A-5

Tables

Introductory table 1. New, modified, and rescinded Army terms vii
Table 2-1. Principles of joint operations 2-1
Table 2-2. Elements of operational art 2-6
Table 3-1. Elements of decisive action 3-2
Table 3-2. Consolidate gains by echelon 3-6
Table 3-3. The Soldier's Rules 3-11
Table A-1. Joint support categories A-7
Table A-2. Command relationships A-10
Table A-3. Army support relationships A-11
Table A-4. Other relationships A-12

Preface

ADP 3-0, *Operations*, constitutes the Army's view of how to conduct prompt and sustained operations across multiple domains, and it sets the foundation for developing other principles, tactics, techniques, and procedures detailed in subordinate doctrine publications. It articulates the Army's operational doctrine for unified land operations. ADP 3-0 accounts for the uncertainty of operations and recognizes that a military operation is a human undertaking. Additionally, this publication is the foundation for training and Army education system curricula related to unified land operations.

The principal audience for ADP 3-0 is all members of the profession of arms. Commanders and staffs of Army headquarters serving as joint task force (JTF) or multinational headquarters should also refer to applicable joint or multinational doctrine concerning the range of military operations and joint or multinational forces. Trainers and educators throughout the Army will use this publication as well.

Commanders, staffs, and subordinates ensure that their decisions and actions comply with applicable United States, international, and in some cases host-nation laws and regulations. Commanders at all levels ensure that their Service members operate in accordance with the law of war and the rules of engagement. (See FM 27-10.)

ADP 3-0 uses joint terms where applicable. Selected joint and Army terms and definitions appear in both the glossary and the text. Terms for which ADP 3-0 is the proponent publication (the authority) are marked with an asterisk (*) in the glossary. Definitions for which ADP 3-0 is the proponent publication are boldfaced in the text. For other definitions shown in the text, the term is italicized and the number of the proponent publication follows the definition.

ADP 3-0 applies to the Active Army, Army National Guard/Army National Guard of the United States, and United States Army Reserve unless otherwise stated.

The proponent of ADP 3-0 is the United States Army Combined Arms Center. The preparing agency is the Combined Arms Doctrine Directorate, United States Army Combined Arms Center. Send comments and recommendations on a DA Form 2028 (*Recommended Changes to Publications and Blank Forms*) to Commander, United States Army Combined Arms Center, Fort Leavenworth, ATTN: ATZL-MCD (ADP 3-0), 300 McPherson Avenue, Fort Leavenworth, KS 66027-2337; by email to usarmy.leavenworth.mccoe.mbx.cadd-org-mailbox@mail.mil; or submit an electronic DA Form 2028.

This page intentionally left blank.

Introduction

ADP 3-0 describes how the Army conducts operations as a unified action partner using the Army's *operational concept*—a fundamental statement that frames how Army forces, operating as part of a joint force, conduct operations (ADP 1-01). The Army's operational concept is unified land operations. ADP 3-0 discusses the foundations, tenets, and doctrine of unified land operations, which serves as a common reference for solving military problems in multiple domains and the framework for the range of military operations across the competition continuum. It is the core of Army doctrine, and it guides how Army forces contribute to unified action. (See the introductory figure on page vi for the ADP 3-0 logic chart.)

ADP 3-0 lists key ideas, such as principles and tenets, as a means of organizing ways to think about military problem solving. A narrative discussion follows each list to provide explanation and context about the subject. The proper application of principles and tenets to a particular situation requires situational understanding informed by professional judgment. Like all doctrine, ADP 3-0 provides a common approach to problem solving, not a list of solutions that can substitute for thinking by commanders and staffs.

ADP 3-0 modifies key topics and updates terminology and concepts as necessary. These topics include the discussion of an operational environment and the operational and mission variables, as well as discussions of unified action, law of land warfare, and combat power. ADP 3-0 maintains combined arms as the application of arms that multiplies Army forces' effectiveness in all operations. However, it expands combined arms to include joint and multinational capabilities as integral to combined arms and discusses how the Army conducts these operations across multiple domains. (For more detailed information on specific tactics and procedures, see FM 3-0.)

ADP 3-0 contains five chapters and one appendix:

Chapter 1 defines military operations, in context, for the Army. It describes the variables that shape the nature of an operational environment and affect outcomes. It provides explanation of unified action and joint operations as well as land operations and the Army's four strategic roles. Finally, it discusses the importance of training to gain skill in land warfare.

Chapter 2 is a discussion on the application of operational art. It details how commanders should consider defeat and stability mechanisms when developing an operational approach. It presents the elements of operational art and describes their meaning.

Chapter 3 addresses the Army's operational concept of unified land operations. It describes how commanders will likely apply landpower as part of unified action to defeat enemy forces on land and establish conditions that accomplish the joint force commander's (JFC's) objectives. Chapter 3 defines the principles and tenets of unified land operations.

Chapter 4 provides the operations structure commanders use to array forces and conduct operations. It also includes the operational framework used in the conduct of unified land operations.

Chapter 5 defines combat power. It discusses the elements of combat power and describes the six warfighting functions used to generate combat power. Lastly, it discusses how Army forces organize combat power through force tailoring, task organization, and mutual support.

Appendix A addresses command and support relationships. It describes these as the basis for unity of command and unity of effort in operations. It details how command relationships and authorities affect Army force generation, force tailoring, and task organization. It further discusses how commanders use Army support relationships when task-organizing Army forces.

Introduction

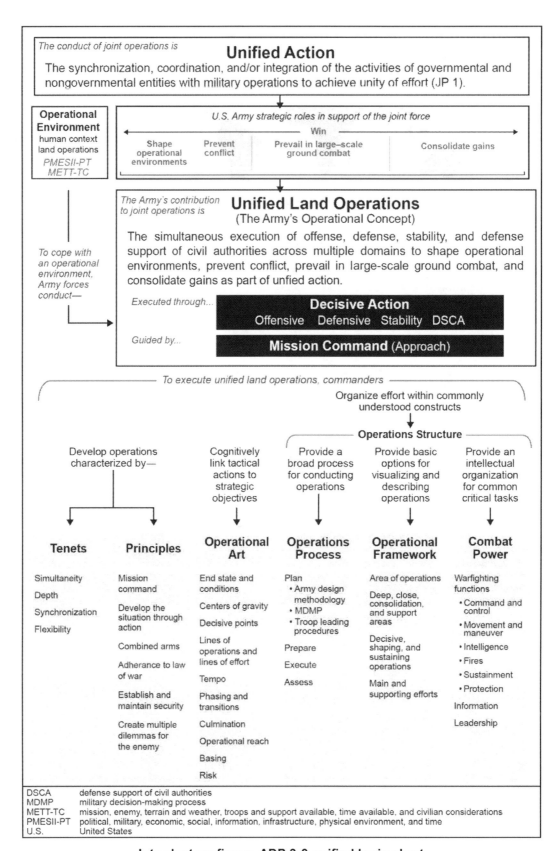

Introductory figure. ADP 3-0 unified logic chart

Certain terms for which ADP 3-0 is the proponent have been added, rescinded, or modified. The glossary contains acronyms and defined terms. (See introductory table 1 for new, modified, and rescinded Army terms.)

Introductory table 1. New, modified, and rescinded Army terms

Term	Reasoning
close area	ADP 3-0 is now the proponent of the term.
close combat	ADP 3-0 is now the proponent of the term. Modified for clarity.
combat power	ADP 3-0 is now the proponent of the term.
combined arms	ADP 3-0 is now the proponent of the term. Modified for clarity.
command and control warfighting function	ADP 3-0 creates new term and definition.
consolidate gains	ADP 3-0 is now the proponent of the term.
consolidation area	ADP 3-0 is now the proponent of the term. Modified for clarity.
cyberspace electromagnetic activities	ADP 3-0 is now the proponent of the term.
decisive action	ADP 3-0 is now the proponent of the term. Modified for clarity.
decisive operation	ADP 3-0 is now the proponent of the term.
deep area	ADP 3-0 is now the proponent of the term.
defeat	ADP 3-0 is now the proponent of the term. Modified for clarity.
defeat mechanism	ADP 3-0 is now the proponent of the term.
defensive operation	ADP 3-0 creates new term and definition.
defensive task	No longer a defined term.
depth	ADP 3-0 is now the proponent of the term.
disintegrate	ADP 3-0 is now the proponent of the term.
dislocate	ADP 3-0 is now the proponent of the term.
enemy	ADP 3-0 is now the proponent of the term.
exterior lines	ADP 3-0 is now the proponent of the term.
fires warfighting function	ADP 3-0 is now the proponent of the term. Modified for clarity.
flexibility	ADP 3-0 is now the proponent of the term.
force tailoring	ADP 3-0 is now the proponent of the term.
hybrid threat	ADP 3-0 is now the proponent of the term. Modified for clarity.
intelligence warfighting function	ADP 3-0 is now the proponent of the term.
interior lines	ADP 3-0 is now the proponent of the term.
isolate	ADP 3-0 is now the proponent of the term. Modified for clarity.
landpower	ADP 3-0 is now the proponent of the term.
large-scale ground combat	ADP 3-0 creates new term and definition.
large-scale ground combat operations	ADP 3-0 creates new term and definition.
line of effort	ADP 3-0 is now the proponent of the term.

Introductory table 1. New, modified, and rescinded Army terms (continued)

Term	Reasoning
line of operations	ADP 3-0 is now the proponent of the term.
main effort	ADP 3-0 is now the proponent of the term.
maneuver	ADP 3-0 creates new Army definition.
mission command warfighting function	Rescinded.
movement and maneuver warfighting function	ADP 3-0 is now the proponent of the term.
offensive operation	ADP 3-0 creates new term and definition.
offensive task	No longer a defined term.
operational initiative	ADP 3-0 is now the proponent of the term. Modified for clarity.
phase	ADP 3-0 is now the proponent of the term.
position of relative advantage	ADP 3-0 is now the proponent of the term.
protection warfighting function	ADP 3-0 is now the proponent of the term.
shaping operation	ADP 3-0 is now the proponent of the term. Modified for clarity.
simultaneity	ADP 3-0 is now the proponent of the term.
stability mechanism	ADP 3-0 is now the proponent of the term.
stability operation	ADP 3-0 creates new term and definition.
support area	ADP 3-0 is now the proponent of the term.
supporting distance	ADP 3-0 is now the proponent of the term.
supporting effort	ADP 3-0 is now the proponent of the term.
supporting range	ADP 3-0 is now the proponent of the term.
sustaining operation	ADP 3-0 is now the proponent of the term.
sustainment warfighting function	ADP 3-0 is now the proponent of the term.
task-organizing	ADP 3-0 is now the proponent of the term.
tempo	ADP 3-0 is now the proponent of the term.
threat	ADP 3-0 is now the proponent of the term.
unified action partners	ADP 3-0 is now the proponent of the term.
unified land operations	ADP 3-0 is now the proponent of the term. Modified for clarity.
warfighting function	ADP 3-0 is now the proponent of the term.

Chapter 1

Military Operations

This chapter discusses military operations, their relationship to operational environments, and the threats that exist within them. The chapter then discusses the Army's strategic roles in the context of unified action and joint operations. Lastly, this chapter discusses land operations and training readiness.

AN OPERATIONAL ENVIRONMENT

1-1. An *operational environment* is a composite of the conditions, circumstances, and influences that affect the employment of capabilities and bear on the decisions of the commander (JP 3-0). Commanders at all levels have their own operational environments for their particular operations. An operational environment for any specific operation comprises more than the interacting variables that exist within a specific physical area. It also involves interconnected influences from the global or regional perspective (for example, politics and economics) that impact on conditions and operations there. Thus, each commander's operational environment is part of a higher echelon commander's operational environment.

1-2. Operational environments include considerations at the strategic, operational, and tactical levels of warfare. At the strategic level, leaders develop an idea or set of ideas for employing the instruments of national power (diplomatic, informational, military, and economic) in a synchronized and integrated fashion to accomplish national objectives. The operational level links the tactical employment of forces to national and military strategic objectives, with the focus being on the design, planning, and conduct of operations using operational art. (See chapter 2 for a discussion of operational art.) The tactical level of warfare involves the employment and ordered arrangement of forces in relation to each other. The levels of warfare help commanders visualize a logical arrangement of forces, allocate resources, and assign tasks based on a strategic purpose, informed by the conditions within their operational environments.

1-3. Broad trends such as globalization, urbanization, technological advances, and failing states affect land operations. These trends can create instability and contribute to an environment of persistent competition and conflict. Persistent conflict is the protracted confrontation among state, nonstate, and individual actors willing to use violence to achieve political and ideological ends. In such an operational environment, commanders must seek and exploit opportunities for success. To exploit opportunities, commanders must thoroughly understand the dynamic nature of their operational environment. Previous experience within a similar operational environment is not enough to guarantee future mission success in the current one.

1-4. Threats seek to employ capabilities to create effects in multiple domains and the information environment to counter U.S. interests and impede friendly operations. Their activities in the information environment, space, and cyberspace attempt to influence U.S. decision makers and disrupt friendly deployment of forces. Land-based threats will attempt to impede joint force freedom of action across the air, land, maritime, space, and cyberspace domains. They will disrupt the electromagnetic spectrum, sow confusion in the information environment, and challenge the legitimacy of U.S. actions. Understanding how threats can present multiple dilemmas to Army forces in all domains helps Army commanders identify (or create), seize, and exploit their own opportunities.

1-5. Some peer threats have nuclear and chemical weapons capabilities and the ability to employ such weapons in certain situations. However, capability does not always equal intent to use, and it is generally presumed that most would use restraint. Preparation and planning that takes nuclear and chemical weapons capabilities into account is of paramount importance in any confrontation with an adversary armed with them. Understanding threat nuclear and chemical weapons doctrine is important, particularly during large-scale ground combat operations.

Chapter 1

1-6. Nuclear terrorism remains a threat to the United States and to international security and stability. Preventing the illicit acquisition of a nuclear weapon, nuclear materials, or related technology and expertise by a violent extremist organization is a significant U.S. national security priority. The more states—particularly rogue states—that possess nuclear weapons or the materials, technology, and knowledge required to make them, the greater the potential risk of terrorist acquisition. Given the nature of terrorist ideologies, commanders and staffs must assume that terrorists would employ a nuclear weapon were they to acquire one.

1-7. Large-scale ground combat operations can occur below the nuclear threshold, and they are not synonymous with total war. **Large-scale ground combat operations are sustained combat operations involving multiple corps and divisions.** Planning for large-scale ground combat operations against enemies possessing nuclear weapons must account for the possibility of their use against friendly forces. The operational approaches employed by joint force commanders (JFCs) may thus be constrained to avoid nuclear escalation in terms of their geographic depth and the assigned objectives. Large-scale ground combat operations, while potentially enormous in scale and scope, are typically limited by the law of war and the political objectives of the conflict itself. Against nuclear armed enemies, the political objectives of a conflict are also informed by the strategic risk inherent in escalation. While the scale and scope of conventional conflict has been smaller than World War II since 1945, it retains its inherent lethality and complexity.

1-8. Successful operations against nuclear and chemically capable peer threats require units prepared to react to the employment of those capabilities and operate degraded in contaminated environments. Planning and training must include active and passive measures for protection against the effects of these weapons, as well as techniques for mitigating their effects to preserve combat power. This includes greater emphasis on dispersion, survivability, and regenerating communications between echelons. These requirements must be incorporated into every facet of doctrine and training, so units and commanders are technically and psychologically prepared for the environment they may encounter. Survivability in this environment should be a training and readiness objective.

1-9. Modern information technology makes the information environment, which includes cyberspace and the electromagnetic spectrum, indispensable to military operations. The *information environment* is the aggregate of individuals, organizations, and systems that collect, process, disseminate, or act on information (JP 3-13). It is a key part of any operational environment, and it will be simultaneously congested and contested during operations. All actors in the information environment—enemy, friendly, or neutral—remain vulnerable to attack by physical, psychological, cyber, or electronic means. (See JP 3-12 for more information on cyberspace operations and the electromagnetic spectrum.)

1-10. No two operational environments are the same. An operational environment consists of many relationships and interactions among interrelated variables. How entities and conditions interact within an operational environment is often difficult to understand and requires continuous analysis. (See paragraphs 1-13 through 1-15 for a discussion of the operational and mission variables.)

1-11. An operational environment continually evolves because of the complexity of human interaction and how people learn and adapt. People's actions change that environment. Some changes can be anticipated, while others cannot. Some changes are immediate and apparent, while other changes evolve over time or are extremely difficult to detect.

1-12. The complex and dynamic nature of an operational environment makes determining the relationship between cause and effect difficult and contributes to the friction and uncertainty inherent in military operations. Commanders must continually assess their operational environments and re-assess their assumptions. Commanders and staffs use the Army design methodology, operational variables, and mission variables to analyze an operational environment to support the operations process. (See paragraphs 4-7 through 4-8 for a discussion of the Army design methodology.)

OPERATIONAL AND MISSION VARIABLES

1-13. An operational environment evolves as each operation progresses. Army leaders use operational variables to analyze and understand a specific operational environment, and they use mission variables to focus on specific elements during mission analysis. (See appendix A to FM 6-0 for a detailed discussion of operational and mission variables.)

Operational Variables

1-14. Army planners describe conditions of an operational environment in terms of operational variables. Operational variables are those aspects of an operational environment, both military and nonmilitary, that may differ from one operational area to another and affect operations. Operational variables describe not only the military aspects of an operational environment, but also the population's influence on it. Using Army design methodology, planners analyze an operational environment in terms of eight interrelated operational variables: political, military, economic, social, information, infrastructure, physical environment, and time (known as PMESII-PT). As soon as a commander and staff have an indication of where their unit will conduct operations, they begin analyzing the operational variables associated with that location. They continue to refine and update that analysis throughout the course of operations.

Mission Variables

1-15. Upon receipt of an order, Army leaders filter information from operational variables into mission variables during mission analysis. They use the mission variables to refine their understanding of the situation. The mission variables consist of mission, enemy, terrain and weather, troops and support available, time available, and civil considerations (METT-TC). Incorporating the analysis of the operational variables with METT-TC ensures that Army leaders consider the best available information about the mission.

THREATS AND HAZARDS

1-16. For every operation, threats are a fundamental part of an operational environment. A ***threat* is any combination of actors, entities, or forces that have the capability and intent to harm United States forces, United States national interests, or the homeland.** Threats may include individuals, organized or unorganized groups, paramilitary or military forces, nation-states, or national alliances. Commanders and staffs must understand how current and potential threats organize, equip, train, employ, and control their forces. They must continually identify, monitor, and assess threats as they adapt and change over time.

1-17. In general, the various actors in any operational area can qualify as an enemy, an adversary, a neutral, or a friend. An ***enemy* is a party identified as hostile against which the use of force is authorized**. An enemy is also called a combatant and is treated as such under the law of war. Enemies will apply advanced technologies (such a cyberspace attack) as well as simple and dual-use technologies (such as improvised explosive devices). Enemies avoid U.S. strengths (such as long-range surveillance and precision strike missiles) through countermeasures (such as integrated air defense systems, dispersion, concealment, and intermingling with civilian populations).

1-18. An *adversary* is a party acknowledged as potentially hostile to a friendly party and against which the use of force may be envisaged (JP 3-0). During competition and conflict, a neutral is an identity applied to a party whose characteristics, behavior, origin, or nationality indicate that it is neither supporting nor opposing friendly forces. Finally, a friend is a party that positively supports U.S. efforts. Land operations often prove complex because enemies, adversaries, neutrals, and friends may physically intermix, often with no easy means to distinguish one from another.

1-19. The term hybrid threat captures the complexity of operational environments, the multiplicity of actors involved, and the blurring of traditionally regulated elements of conflict. **A *hybrid threat* is the diverse and dynamic combination of regular forces, irregular forces, terrorists, or criminal elements acting in concert to achieve mutually benefitting effects**. Hybrid threats combine traditional forces governed by law, military tradition, and custom with unregulated forces that act without constraints on the use of violence. These may involve nation-states using proxy forces or nonstate actors such as criminal and terrorist organizations that employ sophisticated capabilities traditionally associated with states. Hybrid threats are most effective when they exploit friendly constraints, capability gaps, and lack of situational awareness.

1-20. A *hazard* is a condition with the potential to cause injury, illness, or death of personnel; damage to or loss of equipment or property; or mission degradation (JP 3-33). Hazards include disease, extreme weather phenomena, solar flares, and areas contaminated by toxic materials. Hazards can damage or destroy resources, reduce combat power, and contribute to early culmination that prevents mission accomplishment. Understanding hazards and their effects on operations is generally done in the context of terrain, weather, and various other factors related to a particular mission.

1-21. A peer threat is an adversary or enemy able to effectively oppose U.S. forces world-wide while enjoying a position of relative advantage in a specific region. These threats can generate equal or temporarily superior combat power in geographical proximity to a conflict area with U.S. forces. A peer threat may also have a cultural affinity to specific regions, providing them relative advantages in terms of time, space, and sanctuary. They generate tactical, operational, and strategic challenges an order of magnitude more challenging militarily than other adversaries.

1-22. Peer threats can employ resources across multiple domains to create lethal and nonlethal effects with operational significance throughout an operational environment. They seek to delay deployment of U.S. forces and inflict significant damage across multiple domains in a short period to achieve their goals before culminating. A peer threat uses various methods to employ their instruments of power to render U.S. military power irrelevant. Five broad methods, used in combination by peer threats, include—

- Information warfare.
- Preclusion.
- Isolation.
- Sanctuary.
- Systems warfare.

1-23. Enemies and adversaries pursue anti-access and area-denial capabilities, putting U.S. power projection at risk and enabling an extension of their coercive power well beyond their borders. As a result, the United States may be unable to employ forces with complete freedom of action. The ability of U.S. forces to deliberately build up combat power, perform detailed rehearsals and integration activities, and then conduct operations on their own initiative will likely be significantly challenged. Threats might use cyberspace attack capabilities (such as disruptive and destructive malware), electronic warfare, and space capabilities (such as anti-satellite weapons) to disrupt U.S. communications; positioning, navigation, and timing; synchronization; and freedom of maneuver. Finally, enemies may attempt to strike installations outside the continental United States to disrupt or delay deployment of forces. These types of threats are not specific to any single theater of operations, since they have few geographic constraints.

1-24. When dealing with nuclear powered adversaries, the JFC may face constraints to mitigate risk of escalation. Tensions may heighten when employing ground forces that will operate close to an enemy's border or when exploiting offensive success in ways that threaten the viability of an enemy government to maintain power. Because of the potential for nuclear escalation, Army commanders and staffs should consider tensions and the overall strategic situation as they develop operational approaches at their particular echelon.

1-25. Violent extremist organizations work to undermine regional security in areas such as the Middle East and North Africa. Such groups radicalize populations, incite violence, and employ terror to impose their visions on fragile societies. They are strongest where governments are weakest, exploiting people trapped in fragile or failed states. Violent extremist organizations often coexist with criminal organizations, where both profit from illicit trade and the spread of corruption, further undermining security and stability.

WAR AS A HUMAN ENDEAVOR

1-26. War is chaotic, lethal, and a fundamentally human endeavor. It is a clash of wills fought among and between people. All war is inherently about changing human behavior, with each side trying to alter the behavior of the other by force of arms. Success requires the ability to out think an opponent and ruthlessly exploit the opportunities that come from positions of relative advantage. The side that best understands an operational environment adapts more rapidly and decides to act more quickly in conditions of uncertainty is the one most likely to win.

1-27. War is inextricably tied to the populations inhabiting the land domain. All military capabilities are ultimately linked to land and, in most cases, the ability to prevail in ground combat becomes a decisive factor in breaking an enemy's will. Understanding the human context that enables the enemy's will, which includes culture, economics, and history, is as important as understanding the enemy's military capabilities. Commanders cannot presume that superior military capability alone creates the desired effects on an enemy.

Commanders must continually assess whether their operations are influencing enemies and populations, eroding the enemy's will, and achieving the commanders' intended purpose.

1-28. When unified land operations occur among populations, they influence and are influenced by those populations. The results of these interactions are often unpredictable—and potentially uncontrollable. Commanders should seek to do less harm than good to gain the support of populations and, when possible, to influence their behaviors. Gaining support requires a combination of both coercion and incentives, the exact mix of which is unique to each case. During operations to shape operational environments and prevent conflicts, the scale is weighted heavily towards incentivizing desired behavior. However, in large-scale combat operations, coercion may play a larger role. **Large-scale combat operations are extensive joint combat operations in terms of scope and size of forces committed, conducted as a campaign aimed at achieving operational and strategic objectives.** Consolidating gains requires a more balanced approach. Regardless of the context, U.S. forces always operate consistently with international law and their rules of engagement.

1-29. U.S. military forces operate to achieve the goals and accomplish the objectives assigned to them by the President and Secretary of Defense. Normally, these goals and objectives involve establishing security conditions favorable to U.S. interests. Army forces do this as a function of unified action.

ARMY STRATEGIC ROLES

1-30. The Army's primary mission is to organize, train, and equip its forces to conduct prompt and sustained land combat to defeat enemy ground forces and seize, occupy, and defend land areas. The Army accomplishes its mission by supporting the joint force and unified action partners in four strategic roles: shape operational environments, prevent conflict, prevail in large-scale ground combat, and consolidate gains. The strategic roles clarify the enduring reasons for which the Army is organized, trained, and equipped. Strategic roles are not tasks assigned to subordinate units.

SHAPE OPERATIONAL ENVIRONMENTS

1-31. Army operations to shape bring together all the activities intended to promote regional stability and to set conditions for a favorable outcome in the event of a military confrontation. Army operations to shape help dissuade adversary activities designed to achieve regional goals short of military conflict. As part of operations to shape, the Army provides trained and ready forces to geographic combatant commanders to support their combatant command campaign plan. The theater army and subordinate Army forces help the geographic combatant commander in building partner capacity and capability while promoting stability across an area of responsibility. Army operations to shape are continuous throughout a geographic combatant commander's area of responsibility and occur before, during, and after a joint operation within an operational area.

1-32. Shaping activities include security cooperation and forward presence to promote U.S. interests, developing allied and friendly military capabilities for self-defense and multinational operations, continuously setting the theater for operations and training while providing U.S. forces with peacetime and contingency access to a host nation. Regionally aligned and engaged Army forces are essential to accomplishing objectives to strengthen the global network of multinational partners and preventing conflict. The Army garrisons forces and pre-positions equipment in areas to allow national leaders to respond quickly to contingencies. Operational readiness, training, and planning for potential operations by Army forces represent home station activities that are part of operations to shape.

PREVENT CONFLICT

1-33. Army operations to prevent include all activities to deter undesirable actions by an adversary. These operations are typically in response to indications and warnings that an adversary intends to take military action counter to U.S. interests, or in response to adversary activities that are ongoing. They are intended to change an adversary's risk calculus by raising the adversary's costs for actions that threaten U.S. interests. Prevent activities are generally weighted toward actions to protect friendly forces, assets, and partners, and to indicate U.S. intent to execute subsequent phases of a planned operation. As part of a joint force, Army forces may have a significant role in the execution of directed flexible deterrent options and flexible response

Chapter 1

options. Army prevent activities may include mobilization, force tailoring, and other predeployment activities. They may also involve initial deployment into a theater of operations, including echeloning command posts, employing intelligence collection assets, and further developing communications, sustainment, and protection infrastructure to support the JFC's concept of operations. Regardless of the methods used to raise the potential cost for an adversary, the primary deterrent to conflict is the demonstrated ability of a properly manned, equipped, and trained joint force to prevail in large-scale combat.

PREVAIL IN LARGE-SCALE GROUND COMBAT

1-34. During large-scale ground combat operations, Army forces focus on the defeat and destruction of enemy ground forces as part of the joint team. Army forces close with and destroy enemy forces in any terrain, exploit success, and break the opponent's will to resist. Army forces attack, defend, perform stability tasks, and consolidate gains to accomplish national objectives. Divisions and corps are the formations central to the conduct of large-scale combat operations. The ability to prevail in ground combat is a decisive factor in breaking an enemy's capability and will to continue a conflict. Conflict resolution requires the Army to conduct sustained operations with unified action partners as long as necessary to accomplish national objectives.

CONSOLIDATE GAINS

1-35. Army operations to *consolidate gains* **are activities to make enduring any temporary operational success and to set the conditions for a sustainable security environment, allowing for a transition of control to other legitimate authorities.** Consolidation of gains is an integral and continuous part of armed conflict, and it is necessary for achieving success across the range of military operations. Army forces deliberately plan to consolidate gains throughout an operation as part of defeating the enemy in detail to accomplish overall political and strategic objectives. Early and effective consolidation activities are a form of exploitation performed while other operations are ongoing, and they enable the achievement of lasting favorable outcomes in the shortest time span. Army forces perform these activities through decisive action with unified action partners. In some instances, Army forces will be in charge of integrating forces and synchronizing activities to consolidate gains. In other situations, Army forces will be in support. Army forces may consolidate gains for a sustained period of time over large land areas. While Army forces consolidate gains throughout an operation, consolidating gains becomes the overall focus of Army forces after large-scale combat operations have concluded.

UNIFIED ACTION

1-36. *Unified action* is the synchronization, coordination, and/or integration of the activities of governmental and nongovernmental entities with military operations to achieve unity of effort (JP 1). *Unity of effort* is coordination and cooperation toward common objectives, even if the participants are not necessarily part of the same command or organization, which is the product of successful unified action (JP 1). *Unified action partners* **are those military forces, governmental and nongovernmental organizations, and elements of the private sector with whom Army forces plan, coordinate, synchronize, and integrate during the conduct of operations.** Military forces play a key role in unified action before, during, and after operations. The Army's contribution to unified action is unified land operations. (See paragraphs 3-1 through 3-2 for a detailed discussion of unified land operations.)

1-37. The Army is the dominant U.S. fighting force in the land domain. Army forces both depend upon and enable the joint force across multiple domains, including air, land, maritime, space, and cyberspace. This mutual interdependence creates powerful synergies and reflects that all operations are combined arms operations, and all combined arms operations are conducted in multiple domains. The Army depends on the other Services for strategic and operational mobility, joint fires, and other key enabling capabilities. The Army supports other Services, combatant commands, and unified action partners with ground-based indirect fires and ballistic missile defense, defensive cyberspace operations, electronic protection, communications, intelligence, rotary-wing aircraft, logistics, and engineering.

1-38. The Army's ability to set and sustain the theater of operations is essential to allowing the joint force freedom of action. The Army establishes, maintains, and defends vital infrastructure. It also provides the JFC with unique capabilities, such as port and airfield opening; logistics; chemical defense; and reception, staging, and onward movement, and integration of forces.

1-39. Interagency coordination is a key part of unified action. *Interagency coordination* is within the context of Department of Defense involvement, the coordination that occurs between elements of Department of Defense, and participating United States Government departments and agencies for the purpose of achieving an objective (JP 3-0). Army forces conduct and participate in interagency coordination using established liaison, personal engagement, and planning processes.

1-40. Unified action may require interorganizational cooperation to build the capacity of unified action partners. *Interorganizational cooperation* is interaction that occurs among elements of the Department of Defense; participating United States Government departments and agencies; state, territorial, local, and tribal agencies; foreign military forces and government agencies; international organizations; nongovernmental organizations; and the private sector (JP 3-08). Building partner capacity helps to secure populations, protects infrastructure, and strengthens institutions as a means of protecting common security interests. Building partner capacity results from comprehensive interorganizational activities, programs, and military-to-military engagements united by a common purpose. The Army integrates capabilities of operating forces and the institutional force to support interorganizational capacity-building efforts, primarily through security cooperation interactions.

1-41. *Security cooperation* is all Department of Defense interactions with foreign security establishments to build security relationships that promote specific United States security interests, develop allied and partner nation military and security capabilities for self-defense and multinational operations, and provide United States forces with peacetime and contingency access to allied and partner nations (JP 3-20). Security cooperation provides the means to build partner capacity and accomplish strategic objectives. These objectives include

- Building defensive and security relationships that promote U.S. security interests.
- Developing capabilities for self-defense and multinational operations.
- Providing U.S. forces with peacetime and contingency access to host nations to increase situational understanding of an operational environment.

1-42. Army forces support the objectives of the combatant commander's campaign plan in accordance with appropriate policy, legal frameworks, and authorities. The plan supports those objectives through security cooperation, specifically those involving security force assistance and foreign internal defense. *Security force assistance* is the Department of Defense activities that support the development of the capacity and capability of foreign security forces and their supporting institutions (JP 3-20). *Foreign internal defense* is participation by civilian and military agencies of a government in any of the action programs taken by another government or other designated organization to free and protect its society from subversion, lawlessness, insurgency, terrorism, and other threats to its security (JP 3-22).

1-43. Security force assistance and foreign internal defense professionalize and develop security partner capacity to enable synchronized sustaining operations. Army security cooperation interactions enable other interorganizational efforts to build partner capacity. Army forces—including special operations forces—advise, assist, train, and equip partner units to develop unit and individual proficiency in security operations. The institutional force advises and trains partner army activities to build institutional capacity for professional education, force generation, and force sustainment. (See FM 3-22 for more information on Army support to security cooperation.)

COOPERATION WITH CIVILIAN ORGANIZATIONS

1-44. When directed, Army forces provide sustainment and security for civilian organizations, since many lack these capabilities. Within the context of interagency coordination, this refers to non-Department of Defense (DOD) agencies of the U.S. Government. Other government agencies include, but are not limited to, Departments of State, Justice, Transportation, and Agriculture.

1-45. An intergovernmental organization is an organization created by a formal agreement between two or more governments on a global, regional, or functional basis to protect and promote national interests shared by member states. Intergovernmental organizations may be established on a global, regional, or functional basis for wide-ranging or narrowly defined purposes. Examples include the United Nations and the European Union.

1-46. A *nongovernmental organization* is a private, self-governing, not-for-profit organization dedicated to alleviating human suffering; and/or promoting education, health care, economic development, environmental protection, human rights, and conflict resolution; and/or encouraging the establishment of democratic institutions and civil society (JP 3-08). Their mission is generally humanitarian and not one concerned with assisting the military in accomplishing its objectives. In some circumstances, nongovernmental organizations (NGOs) may provide humanitarian aid simultaneously to elements of both sides in a conflict. Nevertheless, there are many situations where the interests of Army forces and NGOs overlap.

1-47. A contractor is a person or business operating under a legal agreement to provide products or services for pay. A contractor furnishes supplies and services or performs work at a certain price or rate based on contracted terms. Contracted support includes traditional goods and services support, but it may also include interpreter communications, infrastructure, and other related support. Contractor employees include contractors authorized to accompany the force as a formal part of the force and local national employees who normally have no special legal status. (See ATP 4-10 for more information on contractors.)

1-48. Most civilian organizations are not under military control, nor does the American ambassador or a United Nations commissioner control them. Civilian organizations have different organizational cultures and norms. Some may be willing to work with Army forces; others may not. Civilian organizations may arrive well after military operations have begun, making personal contact and team building essential. Command emphasis on immediate and continuous coordination encourages effective cooperation. Commanders should establish liaison with civilian organizations to integrate their efforts as much as possible with Army and joint operations. Civil affairs units typically establish this liaison. (See FM 3-57 for more information on civil affairs units.)

MULTINATIONAL OPERATIONS

1-49. *Multinational operations* is a collective term to describe military actions conducted by forces of two or more nations, usually undertaken within the structure of a coalition or alliance (JP 3-16). While each nation has its own interests and often participates within the limitations of national caveats, all nations bring value to an operation. Each nation's force has unique capabilities, and each usually contributes to an operation's legitimacy in terms of international or local acceptability. Army forces should anticipate that most operations will be multinational operations and plan accordingly. (See FM 3-16 for more information on multinational operations.)

1-50. An *alliance* is the relationship that results from a formal agreement between two or more nations for broad, long-term objectives that further the common interests of the members (JP 3-0). Military alliances, such as the North Atlantic Treaty Organization (commonly known as NATO), allow partners to establish formal, standard agreements.

1-51. A coalition is an arrangement between two or more nations for common action. Nations usually form coalitions for specific, limited purposes. A coalition action is an action outside the bounds of established alliances, usually in a narrow area of common interest. Army forces may participate in coalition actions under the authority of a United Nations' resolution.

1-52. Multinational operations present challenges and demands. These include cultural and language issues, interoperability challenges, national caveats on the use of respective forces, the sharing of information and intelligence, and the rules of engagement. Commanders analyze the particular requirements of a mission in the context of friendly force capabilities to exploit the multinational force's advantages and compensate for its limitations. Establishing effective liaison with multinational partners is critical to situational awareness.

1-53. Multinational sustainment requires detailed planning and coordination. Normally each nation provides a national support element to sustain its forces. However, integrated multinational sustainment may improve efficiency and effectiveness. When authorized and directed, an Army theater sustainment command can provide logistics and other support to multinational forces. Integrating support requirements of several nations' forces—often spread over considerable distances and across international boundaries—is critical to the success of multinational operations and requires flexibility, patience, and persistence.

JOINT OPERATIONS

1-54. Single Services may perform tasks and missions to support DOD objectives. However, the DOD primarily employs two or more Services (from two military departments) in a single operation across multiple domains, particularly in combat, through joint operations. *Joint operations* are military actions conducted by joint forces and those Service forces employed in specified command relationships with each other, which of themselves, do not establish joint forces (JP 3-0). A *joint force* is a force composed of elements, assigned or attached, of two or more Military Departments operating under a single joint force commander (JP 3-0). Joint operations exploit the advantages of interdependent Service capabilities in multiple domains through unified action. Joint planning integrates military power with other instruments of national power (diplomatic, economic, and informational) to achieve a desired military end state. The *end state* is the set of required conditions that defines achievement of the commander's objectives (JP 3-0). Joint planning connects the strategic end state to the JFC's operational campaign design and ultimately to tactical missions. JFCs use campaigns and major operations to translate their operational-level actions into strategic results. A *campaign* is a series of related operations aimed at achieving strategic and operational objectives within a given time and space (JP 5-0). A *major operation* is a series of tactical actions (battles, engagements, strikes) conducted by combat forces, coordinated in time and place, to achieve strategic or operational objectives in an operational area (JP 3-0). Planning for a campaign is appropriate when the contemplated military operations exceed the scope of a single major operation. Campaigns are always joint operations. Army forces do not conduct campaigns unless they are designated as a joint task force (JTF). However, Army forces contribute to campaigns through the conduct of land operations. (See JP 5-0 for a discussion of campaigns.)

LAND OPERATIONS

1-55. An *operation* is a sequence of tactical actions with a common purpose or unifying theme (JP 1). The Army's primary mission is to organize, train, and equip forces to conduct prompt and sustained land combat operations and perform such other duties, not otherwise assigned by law, as may be prescribed by the President or the Secretary of Defense (as described in Title 10, United States Code). The Army does this through its operational concept of unified land operations. (See paragraphs 3-1 through 3-2 for a detailed discussion of unified land operations.) Army doctrine aligns with joint doctrine, and it is informed by the nature of land operations. Army forces are employed in accordance with the character of the threat and friendly force capabilities. They conduct operations to preserve vital national interests, most important of which are the sovereignty of the homeland and the preservation of the U.S. constitutional form of government. Army forces are prepared to operate across the range of military operations and integrate with unified action partners as part of a larger effort.

1-56. Army forces, with unified action partners, conduct land operations to shape security environments, prevent conflict, prevail in ground combat, and consolidate gains. Army forces provide multiple options for responding to and resolving crises. Army forces defeat enemy forces, control terrain, secure populations, and preserve joint force freedom of action.

1-57. The dynamic interaction among friendly forces, enemy forces, adversaries, neutral parties, and the environment make land operations exceedingly complex. Understanding each of these elements separately is necessary, but not sufficient, to understand their relationships with each other. Understanding the context of dynamic interaction in each case helps determine what constitutes positions of relative advantage. Exploiting positions of relative advantage allows Army forces to defeat adversaries and enemies at least cost.

1-58. Joint doctrine discusses traditional war as a confrontation between nation-states or coalitions of nation-states. This confrontation typically involves small-scale to large-scale, force-on-force military operations in which enemies use various conventional and unconventional military capabilities against each other. Landpower heavily influences the outcome of wars even when it is not the definitive instrument. **Landpower is the ability—by threat, force, or occupation—to gain, sustain, and exploit control over land, resources, and people.** Landpower is the basis of unified land operations. Landpower includes the ability to—

- Protect and defend U.S. national assets and interests.
- Impose the Nation's will on an enemy, by force if necessary.
- Sustain high tempo operations.

- Engage to influence, shape, prevent, and deter in an operational environment.
- Defeat enemy organizations and control terrain.
- Secure populations and consolidate gains.
- Establish and maintain a stable environment that sets the conditions for political and economic development.
- Address the consequences of catastrophic events—both natural and man-made—to restore infrastructure and reestablish civil services.

ARMY FORCES AND EXPEDITIONARY CAPABILITY AND CAMPAIGN QUALITY

1-59. Swift campaigns, however desirable, are the historical exception. Whenever objectives involve controlling populations or dominating terrain, campaign success usually requires employing landpower for protracted periods. The Army's combination of expeditionary capability and campaign quality contributes sustained landpower to support unified action.

1-60. Expeditionary capability describes the ability to promptly deploy combined arms forces on short notice to any location in the world, capable of conducting operations immediately upon arrival. Expeditionary operations are entirely dependent upon joint air and maritime support. *Operational reach* is the distance and duration across which a force can successfully employ military capabilities (JP 3-0). Adequate operational reach is a necessity for forces to conduct decisive action. (See paragraphs 3-3 through 3-36 for a detailed discussion of decisive action.) Extending operational reach is a significant concern for commanders. To achieve a desired end state, forces must possess the necessary operational reach to establish and maintain conditions that define success. Commanders and staffs increase operational reach through deliberate, focused planning—well in advance of operations when possible—and the appropriate sustainment to facilitate endurance.

1-61. Expeditionary capabilities are more than physical attributes; they begin with a mindset that permeates the force. The ability to deploy the right combination of Army forces to the right place at the right time requires unit leadership focused on the training and readiness essential to deploying. Forward deployed units, forward positioned capabilities, and force projection—from anywhere in the world—all contribute to the Army's expeditionary capabilities. Providing JFCs with expeditionary capabilities requires forces organized and equipped to be versatile and rapidly deployable, and able to sustain operations over time.

1-62. Campaign quality describes the Army's ability to sustain operations as long as necessary to achieve success. Campaign quality is an ability to conduct sustaining operations for as long as necessary, adapting to unpredictable and often profound changes in an operational environment as a campaign unfolds. Army forces are organized, trained, and equipped for endurance. They are essential to the JFC for the conduct of campaigns. Campaigning requires a mindset and vision that complements expeditionary requirements. Army leaders understand the effects of protracted land operations on units and adjust the tempo of operations whenever circumstances allow to prolong their effectiveness.

CLOSE COMBAT

1-63. The nature of close combat in land operations is unique. Combatants routinely come face-to-face with one another in large numbers in a wide variety of operational environments comprising all types of terrain. When other means fail to drive enemy forces from their positions, Army forces close with and destroy or capture them. The outcome of battles and engagements depends on the ability of Army forces to close with enemy forces and prevail in close combat. **Close combat is warfare carried out on land in a direct-fire fight, supported by direct and indirect fires and other assets.** Units involved in close combat employ direct fire weapons supported by indirect fire, air-delivered fires, and nonlethal engagement means. Units in close combat defeat or destroy enemy forces and seize and retain ground. Close combat at lower echelons contains many more interactions between friendly and enemy forces than any other form of combat.

1-64. Close combat is most often linked to difficult terrain where enemies seek to negate friendly advantages in technology and weapon capabilities. Urban terrain represents one of the most likely close combat challenges. The complexity of urban terrain and the density of noncombatants reduce the effectiveness of advanced sensors and long-range weapons. Operations in large, densely populated areas require special considerations. From a planning perspective, commanders view cities as both topographic features and a dynamic system of varying operational entities containing hostile forces, local populations, and infrastructure.

1-65. Effective close combat relies on lethality informed by a high degree of situational understanding across multiple domains. The capacity for physical destruction is the foundation of all other military capabilities, and it is building block of military operations. Army formations are organized, equipped, and trained to employ lethal capabilities in a wide range of conditions. The demonstrated lethality of Army forces provides the credibility essential to deterring adversaries and assuring allies and partners.

1-66. An inherent, complementary relationship exists between using lethal force and applying military capabilities for nonlethal purposes. In wartime, each situation requires a different mix of violence and constraint. Lethal and nonlethal actions used together complement each other and create multiple dilemmas for opponents. During operations short of armed conflict, the lethality implicit in Army forces enables their performance of other tasks effectively with minimal adversary interference.

SEIZE, RETAIN, AND EXPLOIT THE OPERATIONAL INITIATIVE

1-67. ***Operational initiative*** **is the setting of tempo and terms of action throughout an operation.** Army forces seize, retain, and exploit operational initiative by forcing the enemy to respond to friendly action. By presenting an enemy force multiple dilemmas across multiple domains, commanders force that enemy to react continuously until driven into an untenable position. Exploiting operational initiative pressures enemy commanders to abandon their preferred options, react to friendly actions, and make mistakes. As enemy forces make mistakes or weaken, friendly forces seize opportunities that create new avenues for exploitation.

1-68. Commanders seize operational initiative by acting across multiple domains simultaneously. Without action, seizing operational initiative is impossible. Faced with an uncertain situation, commanders naturally tend to hesitate and gather more information to reduce uncertainty. Waiting for more information might reduce uncertainty, but it never eliminates it. Waiting for perfect friendly situational awareness and synchronization provides an adaptive enemy force the time to seize or regain operational initiative. Successful commanders manage uncertainty by developing the situation through action, using reconnaissance, surveillance, and other capabilities to identify opportunities across multiple domains that can be exploited.

1-69. Seizing operational initiative means setting and dictating the terms of action throughout an operation. Commanders plan to seize the initiative as early as possible. Effective planning determines where, when, and how that happens. Enemy forces will actively try to retain operational initiative and disrupt friendly plans, so good plans rapidly executed are fundamental to seizing the initiative. During execution, commanders exploit opportunities to attack and deceive enemy command and control elements to prevent their synchronization of combat power and achieve surprise. Seizing the operational initiative usually requires accepting risk. Commanders and staffs assess if they have the initiative and determine how to seize it if they do not. These conditions generally indicate that friendly forces have operational initiative:

- Friendly forces are no longer decisively engaged or threatened with decisive engagement.
- Subordinate commanders are able to mass combat power or concentrate forces at times and places of their choosing.
- Enemy forces no longer offer effective resistance and do not appear capable of reestablishing resistance.
- Friendly forces encounter lighter-than-anticipated enemy resistance or large numbers of prisoners.
- Friendly rates of advance suddenly accelerate or casualty rates suddenly drop.

1-70. Retaining operational initiative requires sustained, relentless pressure on enemy forces. Commanders maintain pressure by synchronizing the warfighting functions to present enemy commanders with continuously changing combinations of combat power at a tempo they cannot effectively counter. Commanders and staffs use information collection assets to identify enemy attempts to regain the initiative. Effective information management to process information quickly is essential for staying inside the enemy's decision-making cycle. Combined with effective planning, information management helps commanders anticipate enemy actions and develop branches, sequels, or adjustments.

Chapter 1

READINESS THROUGH TRAINING

1-71. Training is the most important thing the Army does to prepare for operations. It is the cornerstone of combat readiness and the foundation for successful operations. Effective training must be commander driven, rigorous, realistic, and to the standard and under the conditions that units expect to operate in during combat. Realistic training with limited time and resources demands that commanders focus their unit training efforts to maximize repetitions under varying conditions to build proficiency. Units execute effective individual and collective training based on the Army's principles of training as described in ADP 7-0. Through training and leader development, units achieve the tactical and technical competence that builds confidence and allows them to conduct successful operations across the competition continuum. Achieving this competence requires specific, dedicated training on offensive, defensive, and stability or defense support of civil authorities (DSCA) tasks. Training continues in deployed units to sustain skills and to adapt to changes in an operational environment. (See ADP 7-0 for training doctrine.)

1-72. Army training includes a system of techniques and standards that allows Soldiers and units to determine, acquire, and practice necessary skills. The Army's training system emphasizes experiential practice and learning to build teamwork and cohesion within units. It recognizes that Soldiers ultimately fight for one another and their units. Training instills discipline. It conditions Soldiers to operate within the law of war and rules of engagement. Training prepares unit leaders for the harsh reality of land combat by emphasizing the fluid and disorderly conditions inherent in land operations. Effective training accounts for cyberspace, space, and information-related capabilities that influence the warfighting functions. Well-rounded training includes candid assessments, after action reviews, and applied lessons learned to ensure improved readiness. Adversaries assess the training readiness of Army forces continuously, which is how training helps to shape operational environments. Training creates combat credibility, which contributes to deterrence.

1-73. Regardless of the importance of technological capabilities, success in operations requires Soldiers to accomplish the mission. Demanding operational environments require professional Soldiers and leaders whose character, commitment, and competence represent the foundation of a values-based, trained, and ready Army. Soldiers and leaders adapt and learn while training to perform tasks both individually and collectively. Soldiers and leaders develop the ability to exercise judgment and disciplined initiative under stress. Army leaders and their subordinates must remain—

- Honorable servants of the Nation.
- Competent and committed professionals of character.
- Dedicated to living by and upholding the Army Ethic.
- Able to articulate mission orders to operate within their commander's intent.
- Committed to developing their subordinates and creating shared understanding while building mutual trust and cohesion.
- Courageous enough to accept risk and exercise disciplined initiative while seeking to exploit opportunities within their commander's intent.
- Trained to operate across the range of military operations.
- Able to operate in combined arms teams within unified action and leverage other capabilities in accomplishing their objectives.
- Opportunistic and offensively minded.

1-74. The complexity of integrating all unified action partners into operations demands that Army forces maintain a high degree of proficiency that is difficult to achieve quickly. Leaders at all echelons seek training opportunities involving the Regular Army and Reserve Components, and with unified action partners at home station, at combat training centers, and when deployed. Formations train in contested conditions that emphasize degraded friendly capabilities, reduced time for preparation, and austere expeditionary conditions.

1-75. U.S. responsibilities are global and Army forces prepare to operate in any environment. Because Army forces face diverse threats and mission requirements, commanders adjust their training priorities based on a likely operational environment. As units prepare for deployment, commanders adapt training priorities and conditions to best address tasks required by actual or anticipated operations. The Army as a whole trains to be flexible enough to operate successfully across the range of military operations. Units train to be agile enough to adapt quickly and shift focus across the competition continuum.

This page intentionally left blank.

Chapter 2
Operational Art

This chapter discusses the application and elements of operational art. It also discusses defeat and stability mechanisms.

THE APPLICATION OF OPERATIONAL ART

2-1. *Operational art* is the cognitive approach by commanders and staffs—supported by their skill, knowledge, experience, creativity, and judgment—to develop strategies, campaigns, and operations to organize and employ military forces by integrating ends, ways, and means (JP 3-0). It is the essence of applying skill, experience, and judgment when exercising military command at the operational-level of warfare, and serves two main functions—

- To ensure that military actions are aligned with, and directly support strategy.
- To ensure that tactical actions occur under the most advantageous conditions possible.

2-2. Commanders and their staffs apply operational art throughout all phases of the operations process. Army commanders use operational art, the principles of joint operations, and the elements of operational art to envision how to establish conditions that accomplish their missions and objectives. For Army forces, operational art is the pursuit of strategic objectives, in whole or in part, through the arrangement of tactical actions in time, space, and purpose. Operational art applies to all types and aspects of operations.

2-3. The twelve principles of joint operations represent important factors that affect the conduct of operations across the levels of warfare. (See table 2-1) The principles are broadly applied considerations and their relevance varies in each situation. They are not a checklist. Commanders generally consider all twelve principles, but they may not apply them in the same way in every operation. The principles summarize the characteristics of successful operations throughout history. Their greatest value lies in educating military professionals. While considering the principles of joint operations, commanders determine if or when to deviate from the principles based on the current situation. (See JP 3-0 for a detailed discussion on the principles of joint operations.)

Table 2-1. Principles of joint operations

Objective: Direct every military operation toward a clearly defined, decisive, and achievable goal.
Offensive: Seize, retain, and exploit the initiative.
Mass: Concentrate the effects of combat power at the most advantageous place and time to produce decisive results.
Maneuver: Place the enemy in a position of disadvantage through the flexible application of combat power.
Economy of force: Expend minimum-essential combat power on secondary efforts to allocate the maximum possible combat power on primary efforts.
Unity of command: Ensure unity of effort under one responsible commander for every objective.
Security: Prevent the enemy from acquiring an unexpected advantage.
Surprise: Strike at a time or place or in a manner for which the enemy is unprepared.
Simplicity: Increase the probability that plans and operations will be executed as intended by preparing clear, uncomplicated plans and concise mission orders.
Restraint: Limit collateral damage and prevent the unnecessary use of force.
Perseverance: Ensure the commitment necessary to attain the national strategic end state.
Legitimacy: Maintain legal and moral authority in the conduct of operations.

Chapter 2

2-4. When applying operational art, commanders and staffs ensure a shared understanding of purpose. This requires open, continuous collaboration between commanders at various echelons to define accurately the problems and conditions of an operational environment. Effective collaboration facilitates assessment, fosters critical analysis, and anticipates opportunities and risk.

2-5. Operational art encompasses all levels, from strategic direction to tactical actions. It requires creative vision, broad experience, and a knowledge of capabilities, tactics, and techniques across multiple domains. It is through operational art that commanders translate their operational approach into a concept of operations. A *concept of operations* is a verbal or graphic statement that clearly and concisely expresses what the commander intends to accomplish and how it will be done using available resources (JP 5-0). Commanders then position and maneuver forces to perform tasks that best achieve a desired end state.

2-6. The successful application of operational art relies heavily on the science of operations. Considerations such as movement times, capability ranges, loiter times, consumption rates, available supplies, combat power status, and electromagnetic spectrum management determine whether an operational approach is feasible or not. Many operational approaches prove unhelpful for driving detailed planning because they fail to consider operational realities. The earlier details are integrated into conceptual planning, the better.

2-7. During planning, commanders and their staffs use the Army design methodology to develop an operational approach that informs detailed planning. The *Army design methodology* is a methodology for applying critical and creative thinking to understand, visualize, and describe unfamiliar problems and approaches to solving them (ADP 5-0). By applying the Army design methodology, commanders and staffs gain a shared understanding of the environment, and they can define the problems preventing the desired end state. This differs from mission analysis, since it is not mission specific. (See figure 2-1.) These items enable commanders and staffs using Army design methodology:

- The principles of joint operations and principles of war.
- The tenets of unified land operations.
- The elements of operational art.
- The defeat mechanisms.
- The stability mechanisms.

Operational Art

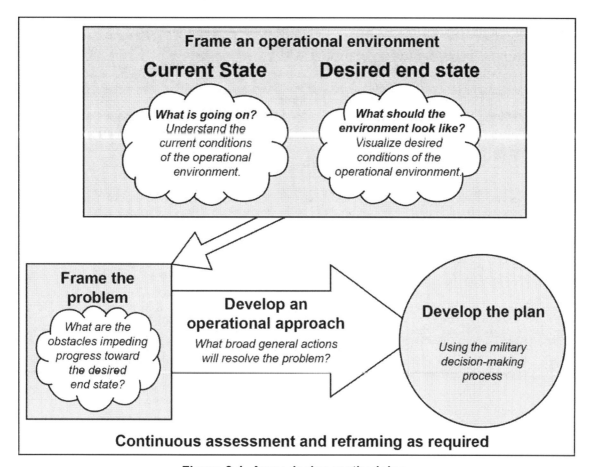

Figure 2-1. Army design methodology

2-8. Army design methodology results in an *operational approach*, a broad description of the mission, operational concepts, tasks, and actions required to accomplish the mission (JP 5-0). A good operational approach provides the basis for detailed planning, allows leaders to establish a logical operational framework, and helps produce an executable order. As detailed planning yields new information, leaders reassess their operational approach—and the Army design methodology that informed it—and adjust it accordingly to ensure relevancy. These actions continue throughout preparation and execution, and they inform commanders' decision-making. When assessing operations, the logic of the operational approach provides the basis for developing assessment criteria, including measures of performance and effectiveness. (See ADP 5-0 for more information on assessments.)

2-9. The understanding developed with the Army design methodology enables commanders to develop an operational approach that establishes conditions to accomplish the mission. (See figure 2-2 on page 2-4.) The operational approach provides a framework that relates tactical tasks to the desired end state. It provides a unifying purpose and focus to all operations.

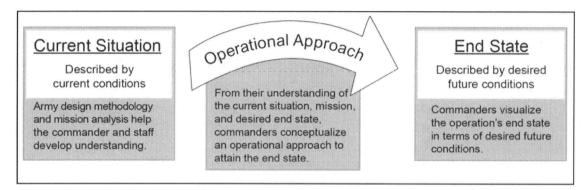

Figure 2-2. Operational approach

DEFEAT AND STABILITY MECHANISMS

2-10. ***Defeat* is to render a force incapable of achieving its objectives**. Defeat has a temporal component and is seldom permanent. When developing an operational approach, commanders consider methods to employ a combination of defeat mechanisms and stability mechanisms. Defeat mechanisms relate to offensive and defensive operations; stability mechanisms relate to stability operations, security, and consolidating gains in an area of operations.

2-11. A ***defeat mechanism* is a method through which friendly forces accomplish their mission against enemy opposition**. Army forces at all echelons use combinations of four defeat mechanisms: destroy, dislocate, disintegrate, and isolate. Applying more than one defeat mechanism simultaneously produces complementary and reinforcing effects not attainable with a single mechanism. Used individually, a defeat mechanism achieves results relative to how much effort is expended. Using defeat mechanisms in combination creates enemy dilemmas that magnify their effects significantly.

2-12. When commanders destroy, they apply lethal combat power on an enemy capability so that it can no longer perform any function. *Destroy* is a tactical mission task that physically renders an enemy force combat-ineffective until it is reconstituted. Alternatively, to destroy a combat system is to damage it so badly that it cannot perform any function or be restored to a usable condition without being entirely rebuilt (FM 3-90-1). An enemy cannot restore a destroyed force to a usable condition without entirely rebuilding it.

2-13. ***Dislocate* is to employ forces to obtain significant positional advantage, rendering the enemy's dispositions less valuable, perhaps even irrelevant**. Commanders often achieve dislocation by placing forces in locations where the enemy does not expect them.

2-14. ***Disintegrate* means to disrupt the enemy's command and control system, degrading its ability to conduct operations while leading to a rapid collapse of the enemy's capabilities or will to fight**. Commanders often achieve disintegration by specifically targeting an enemy's command structure and communications systems.

2-15. ***Isolate* means to separate a force from its sources of support in order to reduce its effectiveness and increase its vulnerability to defeat.** Isolation can encompass multiple domains and can have both physical and psychological effects detrimental to accomplishing a mission. Isolating a force in the electromagnetic spectrum exacerbates the effects of physical isolation by reducing its situational awareness. The ability of an isolated unit to perform its intended mission generally degrades over time, decreasing its ability to interfere with an opposing force's course of action. When commanders isolate, they deny an enemy or adversary access to capabilities that enable an enemy unit to maneuver at will in time and space.

2-16. Commanders describe defeat mechanisms by the three types of effects they produce:
- Physical effects are those things that are material.
- Temporal effects are those that occur at a specific point in time.
- Cognitive effects those that pertain to or affect the mind.

Operational art formulates the most effective, efficient way to apply defeat mechanisms. Physically defeating an enemy deprives enemy forces of the ability to achieve those aims. Temporally defeating an enemy anticipates enemy reactions and counters them before they can become effective. Cognitively defeating an enemy disrupts decision making and deprives that enemy of the will to fight.

2-17. In addition to defeating an enemy, Army forces often seek to stabilize an area of operations by performing stability tasks. Stability tasks are tasks conducted as part of operations outside the United States in coordination with other instruments of national power to maintain or reestablish a safe and secure environment and provide essential governmental services, emergency infrastructure reconstruction, and humanitarian relief. (See ADP 3-0 for more information on stability.) There are six primary stability tasks:

- Establish civil security.
- Establish civil control.
- Restore essential services.
- Support governance.
- Support economic and infrastructure development.
- Conduct security cooperation.

2-18. The combination of stability tasks performed during operations depends on the situation. In some operations, the host nation can meet most or all of the population's requirements. In those cases, Army forces work with and through host-nation authorities. Commanders use civil affairs operations to mitigate how the military presence affects the population and vice versa. Conversely, Army forces operating in a failed state may need to support the local population and work with civilian organizations to restore capabilities. Civil affairs operations are essential in establishing the trust between Army forces and civilian organizations required for effective working relationships.

2-19. A *stability mechanism* **is the primary method through which friendly forces affect civilians in order to attain conditions that support establishing a lasting, stable peace**. As with defeat mechanisms, combinations of stability mechanisms produce complementary and reinforcing effects that accomplish the mission more effectively and efficiently than single mechanisms do alone.

2-20. The four stability mechanisms are compel, control, influence, and support. Compel means to use, or threaten to use, lethal force to establish control and dominance, affect behavioral change, or enforce compliance with mandates, agreements, or civil authority. Control involves imposing civil order. Influence means to alter the opinions, attitudes, and ultimately the behavior of foreign friendly, neutral, adversary, and enemy audiences through messages, presence, and actions. Support establishes, reinforces, or sets the conditions necessary for the instruments of national power to function effectively.

THE ELEMENTS OF OPERATIONAL ART

2-21. In applying operational art, Army commanders and their staffs use intellectual tools to help understand an operational environment and visualize and describe their approach to conducting an operation. Collectively, these tools are the elements of operational art. They help commanders understand, visualize, and describe the integration and synchronization of the elements of combat power and their commander's intent and guidance. Commanders selectively use these tools in any operation. Their broadest application applies to long-term operations.

2-22. Not all elements of operational art apply at all levels of warfare. A company commander concerned about the tempo of an upcoming operation is probably not concerned with an enemies' center of gravity. A corps commander may consider all elements of operational art in developing a plan to support the JFC. As such, the elements of operational art are flexible enough to apply when pertinent.

2-23. As some elements of operational design apply only to JFCs, the Army modifies the elements of operational design into elements of operational art by adding Army-specific elements. During the planning and execution of Army operations, commanders and staffs consider the elements of operational art as they assess the situation. They adjust current and future operations and plans as the operation unfolds, and they reframe as necessary. (See table 2-2 on page 2-6.)

Table 2-2. Elements of operational art

Operational art consists of these elements:	
• End state and conditions. • Center of gravity.* • Decisive points.* • Lines of operations and lines of effort.* • Tempo.	• Phasing and transitions. • Culmination.* • Operational reach.* • Basing.* • Risk.
*Common to elements of operational design.	

END STATE AND CONDITIONS

2-24. The end state is a set of desired future conditions the commander wants to exist when an operation ends. Commanders include the end state in their planning guidance. A clearly defined end state promotes unity of effort; facilitates integration, synchronization, and disciplined initiative; and helps mitigate risk.

2-25. Army operations typically focus on achieving the military end state that may include contributions to establishing nonmilitary conditions. Commanders explicitly describe the end state and its conditions for every operation. Otherwise, missions lack purpose, and operations lose focus. Successful commanders direct every operation toward a clearly defined, conclusive, and attainable end state (the objective). Most military operations require Army forces to consolidate gains to achieve a desired political end state, the exception being a punitive expedition.

2-26. An end state may evolve as an operation progresses. Commanders continuously monitor operations and evaluate their progress. They evaluate the validity of assumptions and running estimates. Commanders use formal and informal assessment methods to assess their progress in achieving an end state and determine if they need to reframe. An end state should anticipate future operations and set conditions for transitions.

CENTER OF GRAVITY

2-27. A *center of gravity* is the source of power that provides moral or physical strength, freedom of action, or will to act (JP 5-0). The loss of a center of gravity can ultimately result in defeat. A center of gravity is an analytical tool for planning operations. It provides a focal point and identifies sources of strength and weakness. However, the concept of center of gravity is only meaningful when considered in relation to the objectives of the mission. Because most enemies represent adaptive, complex systems, they are likely to have multiple centers of gravity. Destroying or capturing one is unlikely to win a campaign or resolve most conflicts.

2-28. Centers of gravity are not limited to military forces, and they can be physical, moral, and virtual. They are part of a dynamic perspective of an operational environment, and they may change as an environment changes. Physical centers of gravity, such as a capital city or military force, are tangible and typically easier to identify, assess, and account for than moral centers of gravity. Physical centers of gravity can often be influenced solely by military means. In contrast, moral centers of gravity are intangible and more difficult to influence. They can include a charismatic leader, powerful ruling elite, or united population. Military means alone usually prove ineffective when targeting moral centers of gravity. Affecting them requires collective, integrated efforts of all instruments of national power. Likewise, a virtual center of gravity may provide the ability to maintain unity of purpose for a disaggregated or decentralized enemy which does not require mutual physical support to accomplish objectives and is not geographically bound. Virtual centers of gravity are usually associated with violent extremist ideologies and organizations, non-nation state actors, or super-empowered individuals, although nation states could also have them.

2-29. A center of gravity has subcomponents comprising a system, such as command and control or logistics, which can be targeted for information collection and attack. This targeting can identify critical vulnerabilities in the system, such as communications or enemy morale, against which commanders can apply friendly capabilities.

2-30. Commanders analyze a center of gravity thoroughly and in detail. Faulty conclusions drawn from hasty or abbreviated analyses can adversely affect operations, waste critical resources, and incur undue risk. Thoroughly understanding an operational environment helps commanders identify and target enemy centers of gravity. This understanding encompasses how enemies organize, fight, and make decisions. It includes their physical and moral strengths and weaknesses. This understanding helps planners identify centers of gravity, their associated decisive points, and the best approach for achieving the desired end state.

DECISIVE POINTS

2-31. A *decisive point* is a geographic place, specific key event, critical factor, or function that, when acted upon, allows commanders to gain a marked advantage over an enemy or contribute materially to achieving success (JP 5-0). Decisive points help commanders select clear, conclusive, attainable objectives that directly contribute to achieving an end state. Geographic decisive points can include port facilities, distribution networks and nodes, and bases of operation. Specific events and elements of an enemy force can be decisive points. Examples of such events include commitment of the enemy operational reserve and reopening a major oil refinery. Space and cyberspace-enabled capabilities may also represent decisive points.

2-32. A common characteristic of decisive points is their importance to centers of gravity. Decisive points are not centers of gravity; they are key to attacking or protecting centers of gravity, and they comprise parts of its system. A decisive point's importance requires an enemy force to commit significant resources to defend it. The loss of a decisive point weakens a center of gravity, and it may expose more decisive points. Identifying and attacking more decisive points can eventually lead to an attack on the center of gravity itself. Commanders identify the decisive points that offer the greatest physical, temporal, or psychological advantage against centers of gravity.

2-33. Decisive points apply to both the operational and tactical levels when shaping the concept of operations. Decisive points enable commanders to seize, retain, or exploit operational initiative. Controlling them is essential to mission accomplishment. Enemy control of a decisive point may stall friendly momentum, force early culmination, or allow an enemy counterattack.

LINES OF OPERATIONS AND LINES OF EFFORT

2-34. Lines of operations and lines of effort link objectives to the end state physically and conceptually. Commanders may describe an operation along lines of operations, lines of effort, or a combination of both. The combination of them may change based on the conditions within an area of operations. Commanders synchronize and sequence actions, deliberately creating complementary and reinforcing effects. The lines then converge on the well-defined, commonly understood end state outlined in the commander's intent.

2-35. Commanders at all levels may use lines of operations and lines of effort to develop tasks and allocate resources. Commanders may designate one line as the decisive operation and others as shaping operations. Commanders synchronize and sequence related actions along multiple lines. Seeing these relationships helps commanders assess progress toward achieving the end state as forces perform tasks and accomplish missions.

Lines of Operations

2-36. A **line of operations** is a line that defines the directional orientation of a force in time and space in relation to the enemy and links the force with its base of operations and objectives. Lines of operations connect a series of decisive points that lead to control of a geographic or force-oriented objective. Operations designed using lines of operations generally consist of a series of actions executed according to a well-defined sequence. A force operates on interior and exterior lines. **Interior lines are lines on which a force operates when its operations diverge from a central point.** Interior lines allow commanders to move quickly against enemy forces along shorter lines of operation. **Exterior lines are lines on which a force operates when its operations converge on the enemy.** Exterior lines allow commanders to concentrate forces against multiple positions on the ground, thus presenting multiple dilemmas to the enemy. Lines of operations tie offensive and defensive operations to the geographic and positional references in the area of operations.

Lines of Effort

2-37. A *line of effort* **is a line that links multiple tasks using the logic of purpose rather than geographical reference to focus efforts toward establishing a desired end state**. Lines of effort are essential to long-term planning when positional references to an enemy or adversary have little relevance. In operations involving many nonmilitary factors, lines of effort may be the only way to link tasks to the end state. Lines of effort are often essential to helping commanders visualize how military capabilities can support the other instruments of national power.

2-38. Commanders use lines of effort to describe their vision of operations to achieve end state conditions. These lines of effort show how individual actions relate to each other and to achieving the end state. Commanders often use stability and DSCA tasks along lines of effort. These tasks link military actions with the broader interagency or interorganizational effort across the levels of warfare. As operations progress, commanders may modify the lines of effort after assessing conditions. Commanders use measures of performance and measures of effectiveness to continually assess operations. A *measure of performance* is an indicator used to measure a friendly action that is tied to measuring task accomplishment (JP 5-0). A *measure of effectiveness* in an indicator used to measure a current system state, with change indicated by comparing multiple observations over time (JP 5-0).

Combining Lines of Operations and Lines of Effort

2-39. Commanders use lines of operations and lines of effort to connect objectives to a central, unifying purpose. The difference between lines of operations and lines of effort is that lines of operations are oriented on physical linkages, while lines of effort are oriented on logical linkages. Combining lines of operations and lines of effort allows a commander to include stability or DSCA tasks in the long-term plan. This combination helps commanders begin consolidating gains and set the end state conditions for transitions in an operation. (See chapter 3 for a discussion of consolidating gains.)

TEMPO

2-40. *Tempo* **is the relative speed and rhythm of military operations over time with respect to the enemy**. It reflects the rate of military action. Controlling tempo helps commanders keep operational initiative during combat operations or rapidly establish a sense of normalcy during humanitarian crises. During combat operations, commanders normally seek to maintain a higher tempo than enemy forces do. A rapid tempo can overwhelm an enemy force's ability to counter friendly actions. During other operations, commanders act quickly to control events and deny enemy forces positions of advantage. By acting faster than the situation deteriorates, commanders can change the dynamics of a crisis and restore favorable conditions.

2-41. Commanders control tempo throughout the conduct of operations. First, they formulate operations that exploit the complementary and reinforcing effects of simultaneous and sequential operations. They synchronize those operations in time and space to degrade enemy capabilities throughout the area of operations. Second, commanders avoid unnecessary engagements. They do this by bypassing resistance and avoiding places not considered decisive. Third, through mission command, commanders enable subordinates to exercise initiative and act independently. Controlling tempo requires both audacity and patience. Audacity initiates the actions needed to develop a situation; patience allows a situation to develop until the force can strike at the most crucial time and place. Ultimately, the goal is maintaining a tempo appropriate to retaining and exploiting the initiative and achieving the end state.

2-42. Army forces expend more energy and resources when operating at a high tempo. Commanders assess their force's capacity to operate at a higher tempo based on its performance and available resources. An effective operational design varies tempo throughout an operation to increase endurance while maintaining appropriate speed and momentum. There is more to tempo than speed. While speed can be important, commanders vary speed to achieve endurance and optimize operational reach.

2-43. When considering tempo it is critical to consider the risks associated with the requirement to consolidate gains. When forces consolidate gains throughout an operation, a commander may accept the risk of slower tempo in the near term, to ensure the enemy is unable to protract the conflict with bypassed or irregular forces that avoid decisive engagement with friendly forces. There may be circumstances when a commander accepts risk by deciding to consolidate gains in a later phase during operations because of a need to conduct operations at a higher tempo initially. Regardless of where the commander accepts risk associated with tempo, the requirement to consolidate gains is inherent to almost all operations on land.

PHASING AND TRANSITIONS

2-44. A *phase* **is a planning and execution tool used to divide an operation in duration or activity**. A change in phase usually involves a change of mission, task organization, or rules of engagement. Phasing helps in planning and controlling, and it may be indicated by time, distance, terrain, or an event. The ability of Army forces to extend operations in time and space, coupled with a desire to dictate tempo, often presents commanders with more objectives and decisive points than the force can engage simultaneously. This may require commanders and staffs to consider sequencing operations.

2-45. Phasing is critical to arranging all tasks of an operation that cannot be performed simultaneously. It describes how the commander envisions the overall operation unfolding. It is the logical expression of the commander's visualization in time. Within a phase, a large portion of the force executes similar or mutually supporting activities. Achieving a specified condition or set of conditions typically marks the end of a phase.

2-46. Simultaneity, depth, and tempo are vital to all operations. However, forces cannot always attain them to the degree desired. In such cases, commanders limit the number of objectives and decisive points engaged simultaneously. They deliberately sequence certain actions to maintain tempo while focusing combat power at a decisive point in time and space. Commanders employ a combination of simultaneous and sequential tasks at multiple echelons to establish end state conditions during an operation.

2-47. Phasing can extend operational reach. When a force lacks the capability to accomplish its mission in a single action, commanders phase the operation. Each phase should strive to—
- Focus effort.
- Concentrate combat power in time and space at a decisive point.
- Accomplish its objectives deliberately and logically.

2-48. Transitions mark a change of focus between phases or between the ongoing operation and execution of a branch or sequel. Shifting priorities among offensive, defensive, stability, and DSCA tasks also involve transitions. Transitions require planning and preparation well before their execution, so the force can maintain the momentum and tempo of operations. The force is vulnerable during transitions, and commanders establish clear conditions for their execution.

2-49. A transition occurs for several reasons. Transitions occur when delivering essential services, retaining infrastructure needed for reconstruction, or when consolidating gains. (See paragraphs 3-28 through 3-38 for a discussion of consolidating gains.) An unexpected change in conditions may require commanders to direct an abrupt transition between phases. In such cases, the overall composition of the force remains unchanged despite sudden changes in mission, task organization, and rules of engagement. Typically, task organization evolves to meet changing conditions; however, transition planning must also account for changes in mission. Commanders continuously assess the situation, and they task-organize and cycle their forces to retain operational initiative. Commanders strive to achieve changes in emphasis without incurring an operational pause.

2-50. Commanders identify potential transitions during planning and account for them throughout execution. Considerations for identifying potential transitions should include—
- Forecasting in advance when and how to transition.
- Arranging tasks to facilitate transitions.
- Creating a task organization that anticipates transitions.
- Rehearsing certain transitions such as from defense to counterattack or from offense to consolidating gains.
- Ensuring the force understands the different rules of engagement during transitions.

2-51. Commanders should appreciate the time required to both plan for and execute transitions. Assessment ensures that commanders measure progress toward such transitions and take appropriate actions to prepare for and execute them.

Chapter 2

CULMINATION

2-52. The *culminating point* is a point at which a force no longer has the capability to continue its form of operations, offense or defense (JP 5-0). Culmination represents a crucial shift in relative combat power. It is relevant to both attackers and defenders at each level of warfare. While conducting offensive operations, the culminating point occurs when a force cannot continue the attack and must assume a defensive posture or execute an operational pause. While conducting a defense, it occurs when a force can no longer defend itself and must withdraw or risk destruction. The culminating point is more difficult to identify when Army forces perform stability tasks. Two conditions can result in culmination while performing stability tasks: units being too dispersed to achieve security and units lacking required resources to achieve the end state. While performing DSCA tasks, culmination may occur if forces must respond to more catastrophic events than they can manage simultaneously. Such a situation results in culmination due to exhaustion.

2-53. A culmination may be a planned event. In such cases, the concept of operations predicts which part of a force will culminate, and the task organization includes additional forces to assume the mission after culmination. Typically, culmination is caused by direct combat actions or higher echelon resourcing decisions. Culmination relates to the force's ability to generate and apply combat power, and it is not a lasting condition. To continue operations after culminating, commanders may reinforce or reconstitute tactical units.

OPERATIONAL REACH

2-54. Operational reach reflects the ability to achieve success through a well-conceived operational approach, and it is applicable to Army forces operating as part of the joint force. Operational reach is a tether; it is a function of intelligence, protection, sustainment, endurance, and combat power relative to enemy forces. The limit of a unit's operational reach is its culminating point. Operational reach balances the natural tension among endurance, momentum, and protection. Commanders seek to extend the operational reach far enough to accomplish their objectives before culmination.

2-55. Endurance refers to the ability to employ combat power anywhere for protracted periods. It stems from the ability to organize, protect, and sustain a force, regardless of the distance from its base and the austerity of the environment. Endurance involves anticipating requirements and making the most effective and efficient use of resources. Endurance contributes to Army forces' ability to achieve decisive outcomes over time.

2-56. Momentum comes from retaining operational initiative and executing high-tempo operations that overwhelm enemy resistance. Commanders control momentum by maintaining focus and pressure. They set a tempo that prevents exhaustion and maintains adequate sustainment. A sustainable tempo extends operational reach. Commanders maintain momentum by anticipating and transitioning rapidly between any combination of offensive, defensive, stability, or DSCA tasks. Momentum prevents an enemy from recovering the initiative. Sometimes commanders push the force to its culminating point to take maximum advantage of an opportunity. Exploitations and pursuits often involve pushing all available forces to the limit of their endurance to capitalize on momentum and retain the initiative.

2-57. Protection is an important contributor to operational reach. Commanders anticipate how enemy actions and environmental factors might disrupt operations and then determine the protection capabilities required to maintain sufficient reach. Protection closely relates to endurance and momentum. It also contributes to the commander's ability to extend operations in time and space. The protection warfighting function helps commanders maintain their force's integrity and combat power.

2-58. Commanders and staffs consider operational reach to ensure Army forces accomplish their missions before culminating. Commanders continually strive to extend operational reach. They assess friendly and enemy force status and civil considerations, anticipate culmination, consolidate gains, and plan operational pauses if necessary. The use of basing can sustain operational reach in time and space.

BASING

2-59. Army basing overseas typically falls into two general categories: permanent (bases or installations) and nonpermanent (base camps). A *base* is a locality from which operations are projected or supported (JP 4-0). Generally, bases are in host nations in which the United States has a long-term lease and a status-of-forces agreement. A *base camp* is an evolving military facility that supports the military operations of a deployed unit and provides the necessary support and services for sustained operations (ATP 3-37.10). Base camps are nonpermanent by design and designated as bases when the intention is to make them permanent. Bases or base camps may have a specific purpose (such as serving as an intermediate staging base, a logistics base, or a base camp) or they may be multifunctional. The longer base camps exist, the more they exhibit many of the same characteristics as bases in terms of the support and services provided and types of facilities developed. A base or base camp has a defined perimeter, has established access controls, and takes advantage of natural and man-made features.

2-60. Basing may be joint or single Service and will routinely support both U.S. and multinational forces, as well as interagency partners, operating anywhere along the range of military operations. Commanders often designate an area as a base or base camp and assign responsibility to a single commander for protection and terrain management within the base. Within large echelon support areas or joint security areas, controlling commanders may designate base clusters for mutual protection and to exercise command and control. (See JP 4-0 for more information on joint logistics and basing and JP 3-10 for more on joint security areas.)

2-61. When a base camp expands to include clusters of sustainment, headquarters, and other supporting units, its commander may designate it a support area. These areas of operations facilitate the positioning, employment, and protection of resources required to sustain, enable, and control tactical operations. Army forces typically rely on a mix of bases and base camps to serve as intermediate staging bases, lodgments (subsequently developed into base camps or potentially bases), and forward operating bases. These bases and base camps deploy and employ landpower simultaneously to operational depth. They establish and maintain strategic reach for deploying forces and ensure sufficient operational reach to extend operations in time and space. (See paragraph 4-29 for a discussion of support areas.)

2-62. An *intermediate staging base* is a tailorable, temporary location used for staging forces, sustainment and/or extraction into and out of an operational area (JP 3-35). At an intermediate staging base, units are unloaded from intertheater lift, reassembled and integrated with their equipment, and then moved by intratheater lift into the area of operations. The theater army commander provides extensive support to Army forces transiting the base. The combatant commander may designate the theater army commander to command the base or provide a headquarters suitable for the task. Intermediate staging bases are established near, but normally not in, the joint operations area. They often are located in the supported combatant commander's area of responsibility. For land forces, intermediate staging bases may be located in the area of operations. However, if possible, they are established outside the range of direct and most indirect enemy fire systems and beyond the enemy's political sphere of influence.

2-63. A base camp that expands to include an airfield may become a forward operating base. A *forward operating base* is an airfield used to support tactical operations without establishing full support facilities (JP 3-09.3). Forward operating bases may be used for an extended time and are often critical to security. During protracted operations, they may be further expanded and improved to establish a more permanent presence. The scale and complexity of a forward operating base, however, directly relate to the size of the force required to maintain it. A large forward operating base with extensive facilities requires a much larger security force than a smaller, austere base. Commanders weigh whether to expand and improve a forward operating base against the type and number of forces available to secure it, the expected length of the forward deployment, the force's sustainment requirements, and the enemy threat.

2-64. A *lodgment* is a designated area in a hostile or potentially hostile operational area that, when seized and held, makes the continuous landing of troops and materiel possible and provides maneuver space for subsequent operations (JP 3-18). Identifying and preparing the initial lodgment significantly influences the conduct of an operation. Lodgments should expand to allow easy access to strategic sealift and airlift, offer adequate space for storage, facilitate transshipment of supplies and equipment, and be accessible to multiple lines of communications. Typically, deploying forces establish lodgments near key points of entry in the operational area that offer central access to air, land, and sea transportation hubs.

RISK

2-65. Risk is the probability and severity of loss linked to hazards. Risk, uncertainty, and chance are inherent in all military operations. When commanders accept risk, they create opportunities to seize, retain, and exploit

Chapter 2

operational initiative and achieve decisive results. The willingness to incur risk is often the key to exposing enemy weaknesses that an enemy considers beyond friendly reach. Understanding risk requires accurate running estimates and valid assumptions. Embracing risk as opportunity requires situational awareness and imagination, as well as audacity. Successful commanders assess and mitigate risk continuously throughout the operations process.

2-66. Inadequate planning and preparation puts forces at risk, as does delaying action while waiting for perfect intelligence and synchronization. Risk averse commanders and units miss fleeting opportunities, which can actually increase the risk of greater casualties. Reasonably estimating and intentionally accepting risk is fundamental to conducting successful operations and essential to the mission command approach. Experienced commanders balance audacity and imagination against risk and uncertainty to strike at a time, at a place, and in a manner unexpected by enemy forces. This is the essence of surprise.

2-67. Commanders accept risks to create and maintain conditions necessary to seize, retain, and exploit the initiative. A good operational approach considers the balances of risk and uncertainty with friction and chance. Plans and orders should provide the flexibility subordinates need take initiative when opportunities present themselves or conditions change as a hedge against risk. Plans and orders that have no tolerances for friction or deviation from how tasks might be accomplished are inherently higher risk than those that do.

Chapter 3
The Army's Operational Concept

This chapter discusses the Army's operational concept of unified land operations. It discusses the principles and tenets of unified land operations and decisive action.

UNIFIED LAND OPERATIONS

3-1. Unified land operations is the Army's warfighting doctrine, and it is the Army's operational concept and contribution to unified action. Unified land operations is an intellectual outgrowth of both previous operations doctrine and recent combat experience. It recognizes the nature of modern warfare in multiple domains and the need to conduct a fluid mix of offensive, defensive, and stability operations or DSCA simultaneously. Unified land operations acknowledges that strategic success requires fully integrating U.S. military operations with the efforts of interagency and multinational partners. Army forces, as part of the joint force, contribute to joint operations through the conduct of unified land operations. **Unified land operations is the simultaneous execution of offense, defense, stability, and defense support of civil authorities across multiple domains to shape operational environments, prevent conflict, prevail in large-scale ground combat, and consolidate gains as part of unified action.**

3-2. The goal of unified land operations is to establish conditions that achieve the JFC's end state by applying landpower as part of a unified action to defeat the enemy. Unified land operations is how the Army applies combat power through 1) simultaneous offensive, defensive, and stability, or DSCA, to 2) seize, retain, and exploit the initiative, and 3) consolidate gains. Military forces seek to prevent or deter threats through unified action, and, when necessary, defeat aggression.

DECISIVE ACTION

3-3. **Decisive action is the continuous, simultaneous execution of offensive, defensive, and stability operations or defense support of civil authority tasks.** Army forces conduct decisive action. Commanders seize, retain, and exploit the initiative while synchronizing their actions to achieve the best effects possible. Operations conducted outside the United States and its territories simultaneously combine three elements of decisive action—offense, defense, and stability. Within the United States and its territories, decisive action combines elements of DSCA and, as required, offense and defense to support homeland defense. (See table 3-1 on page 3-2.)

3-4. Decisive action begins with the commander's intent and concept of operations. Decisive action provides direction for an entire operation. Commanders and staffs refine the concept of operations during planning and determine the proper allocation of resources and tasks. Throughout an operation, they may adjust the allocation of resources and tasks as conditions change.

3-5. The simultaneity of decisive action varies by echelon and span of control. Higher echelons generally have a broader focus than lower echelons when assigning responsibilities to subordinates. The higher the echelon, the greater the possibility that all elements of decisive action occur simultaneously within its area of operations. At lower echelons, an assigned task may require all the echelons' combat power to execute a specific task. For example, in some form a higher echelon, such as a corps, always performs offensive, defensive, and stability or defense support of civil authority operations simultaneously. Subordinate brigades perform some combination of offensive, defensive, and stability operations, but they generally are more focused by their immediate priorities on a specific element, particularly during large-scale ground combat operations.

Chapter 3

Table 3-1. Elements of decisive action

Offense	Defense
Types of Offensive Operations	**Types of Defensive Operations**
• Movement to contact. • Attack. • Exploitation. • Pursuit.	• Mobile defense. • Area defense. • Retrograde.
Purposes	**Purposes**
• Dislocate, isolate, disrupt, and destroy enemy forces. • Seize key terrain. • Deprive the enemy of resources. • Refine intelligence. • Deceive and divert the enemy. • Provide a secure environment for stability tasks.	• Deter or defeat enemy offense. • Gain time. • Achieve economy of force. • Retain key terrain. • Protect the population, critical assets, and infrastructure. • Refine intelligence.
Stability	*Defense Support of Civil Authorities*
Stability Operations Tasks	**Defense Support of Civil Authorities Tasks**
• Establish civil security. • Establish civil control. • Restore essential services. • Support to governance. • Support to economic and infrastructure development. • Conduct security cooperation.	• Provide support for domestic disasters. • Provide support for domestic chemical, biological, radiological, and nuclear incidents. • Provide support for domestic civilian law enforcement agencies. • Provide other designated support.
Purposes	**Purposes**
• Provide a secure environment. • Secure land areas. • Meet the critical needs of the population. • Gain support for host-nation government. • Shape the environment for interagency and host-nation success. • Promote security, build partner capacity, and provide access. • Refine intelligence.	• Save lives. • Restore essential services. • Maintain or restore law and order. • Protect infrastructure and property. • Support maintenance or restoration of local government. • Shape the environment for intergovernmental success.

3-6. Unified land operations addresses combat with armed opponents amid populations. This requires Army forces to shape civil conditions. Winning battles and engagements is important, but it is not always the most significant task in a strategic context. Shaping civil conditions with unified action partners is generally important to the success of all campaigns, and thus it is a critical component of all operations.

3-7. Unified land operations span the entire competition continuum. They are conducted to support all four Army strategic roles. The relative emphasis on the various elements of decisive action vary with the purpose and context of the operations being conducted. (See figure 3-1.)

The Army's Operational Concept

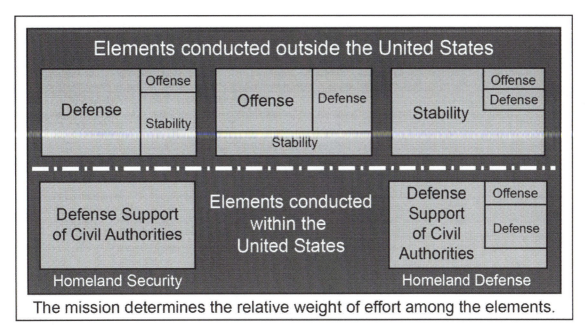

Figure 3-1. Decisive action

3-8. There are enabling operations that are common to all elements of decisive action. Examples of enabling operations are relief in place, security operations, and troop movement. (For more information about enabling operations, see ADP 3-90.)

3-9. Historical experience demonstrates that forces trained exclusively for offensive and defensive operations are not fully proficient at stability operations. Likewise, forces involved in protracted stability operations or DSCA require intensive training to regain proficiency in offensive or defensive operations before engaging in large-scale ground combat operations. While there is inherent risk in either situation, there is greater risk to the force when it is required to conduct offensive and defensive operations for which it is not proficient. Commanders resolve the tension inherent in the different training requirements by setting priorities for individual, collective, and unit mission-essential tasks based upon higher echelon guidance and what their units are most likely to be tasked to do.

THE PURPOSE OF SIMULTANEITY

3-10. Simultaneity is the act of doing multiple things at the same time. The purpose of simultaneity during decisive action is to create multiple dilemmas that overwhelm an adversary or enemy's ability to effectively respond. Multiple dilemmas can create a cascading effect that leaves an opponent with no good options to accomplish its objectives. Achieving simultaneity requires the ability to conduct operations in depth and to integrate them so that their timing multiplies their effectiveness across multiple domains throughout an area of operations. Commanders must consider their entire area of operations, enemy forces, and information collection activities as they synchronize combat power to conduct simultaneous operations that immobilize, suppress, or surprise enemy forces. Such actions nullify an enemy force's ability to react in a coordinated, mutually supporting fashion.

3-11. Army forces create depth in time and space through combined arms, economy of force, continuous reconnaissance, and joint capabilities. Conducting operations across large areas forces an adversary or enemy to react in multiple directions and opens up opportunities that can be further exploited to create additional dilemmas.

Offensive Operations

3-12. An *offensive operation* **is an operation to defeat or destroy enemy forces and gain control of terrain, resources, and population centers.** Offensive operations impose the commander's will on an enemy. The offense is the most direct means of seizing, retaining, and exploiting the initiative to gain a physical and psychological advantage. In the offense, the decisive operation is a sudden action directed toward enemy weaknesses and capitalizing on speed, surprise, and shock. If that operation fails to destroy an enemy, operations continue until enemy forces are defeated. The offense compels an enemy to react, creating new or larger weaknesses the attacking force can exploit. (See ADP 3-90 for a detailed discussion of the offense.)

Defensive Operations

3-13. A *defensive operation* **is an operation to defeat an enemy attack, gain time, economize forces, and develop conditions favorable for offensive or stability operations.** Normally the defense cannot achieve a decisive victory. However, it sets conditions for a counteroffensive or a counterattack that enables forces to regain the initiative. Defensive operations are a counter to an enemy offensive action, and they seek to destroy as much of the attacking enemy forces as possible. They preserve control over land, resources, and populations, and retain key terrain, protect lines of communications, and protect critical capabilities against attack. Commanders can conduct defensive operations in one area to free forces for offensive operations elsewhere. (See ADP 3-90 for a detailed discussion of the defense.)

Stability Operations

3-14. A *stability operation* **is an operation conducted outside the United States in coordination with other instruments of national power to establish or maintain a secure environment and provide essential governmental services, emergency infrastructure reconstruction, and humanitarian relief.** These operations support governance by a host nation, an interim government, or a military government. Stability involves coercive and constructive action. Stability helps in building relationships among unified action partners and promoting U.S. security interests. It can help establish political, legal, social, and economic institutions in an area while supporting transition of responsibility to a legitimate authority. Commanders are legally required to perform minimum-essential stability operations tasks when controlling populated areas of operations. These include security, food, water, shelter, and medical treatment. (See ADP 3-07 for a detailed discussion of stability.)

Defense Support of Civil Authorities

3-15. *Defense support of civil authorities* is support provided by U.S. Federal military forces, DOD civilians, DOD contract personnel, DOD Component assets, and National Guard forces (when the Secretary of Defense, in coordination with the Governors of the affected States, elects and requests to use those forces in Title 32, United States Code status) in response to requests for assistance from civil authorities for domestic emergencies, law enforcement support, and other domestic activities, or from qualifying entities for special events. (DODD 3025.18). DSCA is a task executed in the homeland and U.S. territories. It is performed to support another primary agency, lead federal agency, or local authority. When DSCA is authorized, it consists of four types of operations (see table 3-1 on page 3-2 and DODD 3025.18). National Guard forces—Title 32 or state active forces under the command and control of the governor and the adjutant general—are usually the first forces to respond on behalf of state authorities. When Federal military forces are employed for DSCA activities, they remain under Federal military command and control at all times. (See DODD 3025.18, JP 3-28, and ADP 3-28 for detailed discussions of DSCA.)

HOMELAND DEFENSE AND DECISIVE ACTION

3-16. *Homeland defense* is the protection of United States sovereignty, territory, domestic population, and critical infrastructure against external threats and aggression or other threats as directed by the President (JP 3-27). The DOD has lead responsibility for homeland defense. The strategy for homeland defense (and DSCA) calls for defending U.S. territory against attack by state and nonstate actors through an active, layered defense that aims to deter and defeat aggression abroad and simultaneously protect the homeland. The Army supports this strategy with capabilities in forward regions of the world, geographic approaches to U.S. territory, and within the U.S. homeland.

3-17. During homeland defense, Army forces work closely with federal, state, territorial, tribal, local, and private agencies. Land domain homeland defense could consist of offense and defense as part of decisive action. Homeland defense is a defense-in-depth that relies on collection, analysis, and sharing of information and intelligence; strategic and regional deterrence; military presence in forward regions; and the ability to rapidly obtain and project warfighting capabilities to defend the United States, its allies, and its interests. This defense may include support to civil law enforcement; antiterrorism and force protection; counterdrug; air and missile defense; chemical, biological, radiological, nuclear, and high-yield explosives; and defensive cyberspace operations. It can also include security cooperation with other partners to build an integrated and mutually supportive concept of protection.

TRANSITIONING IN DECISIVE ACTION

3-18. Conducting decisive action involves more than simultaneous execution. Commanders and staffs must consider their units' capabilities and capacities relative to each element of decisive action. Commanders consider and balance the elements while preparing their commander's intent and concept of operations. They determine which elements to accomplish simultaneously and which require phasing, whether additional resources are necessary, and how to transition emphasis from one to another.

3-19. Transitions in emphasis between the elements of decisive action require careful assessment, planning, and unit preparation. Commanders first assess a situation to determine applicable elements and the priority for each. When an operation is phased, the plan includes these changes. The relative weight given to each element varies with conditions. This weight is reflected in tasks assigned to subordinates, resource allocation, and task organization.

3-20. Unanticipated changes or an improved understanding of an operational environment may result in commanders reframing a problem and adapting an operation. Unforeseen success resulting in collapse of enemy opposition illustrates one unanticipated change. Another example is a deteriorating situation during peace operations requiring a transition to the defense or offense to reestablish stability. Commanders need to adjust task organizations to support the requirements of transitions. When transitioning, subordinate commanders must clearly understand their higher echelon commander's intent, concept of operations, and desired end state. This includes how much risk to accept, and where to accept it.

CONSOLIDATE GAINS

3-21. The Army strategic role of consolidate gains sets conditions for enduring political and strategic outcomes to military operations. Army forces provide most of the capabilities the JFC requires to consolidate gains at scale during a campaign. Army and unified action partner forces exploit tactical and operational success for the JFC as they consolidate gains to set security conditions for the desired political end state. Activities to consolidate gains are an integral part of winning across the competition continuum and range of military operations, and they require consideration through all phases of an operation. Determining when and how to consolidate gains at the operational level, and applying the necessary resources at the tactical level to do so effectively, requires clear understanding about where to accept risk during an operation. Failure to consolidate gains generally leads to failure in achieving the desired end state, since it would represent a failure to follow-through on initial tactical successes and cedes the initiative to determined enemies seeking to prolong a conflict. The creation of governable space is necessary for transition of responsibility to a legitimate authority and the successful completion of combat operations. Army forces integrate the efforts of all unified action partners as they consolidate gains.

3-22. Army forces consolidate gains through decisive action, executing offense, defense, and stability to defeat enemy forces in detail and set security conditions required for a desired end state. Consolidate gains is not a phase. Army forces consolidate gains continuously during the conduct of operations, although not simultaneously and with varying purposes by echelons over time. Consolidating gains is focused on the exploitation of tactical success to ensure enemy forces cannot reconstitute any form of resistance in areas where they were initially defeated. This creates an enabling tempo of operations on the ground in the close, deep, and support areas. (See table 3-2 on page 3-6 for a general taxonomy of purpose that reflects focus, planning considerations, and approach by echelons as they consolidate gains during combat.)

3-23. At the strategic-operational level the highest tactical echelons plan and coordinate the resources necessary to achieve the JFC's desired end state. They also provide subordinate echelons a shared visualization of the security conditions necessary for the desired political or strategic end state. Achieving

Chapter 3

the desired end state generally requires a whole of government effort with unified action partners in and out of the theater of operations. At the operational-tactical level, field armies and corps exploit division tactical success through decisive action by maintaining contact with enemy remnants, bypassed forces, and the capabilities that enemy forces could militarize to protract the conflict. Friendly forces employ lethal and non-lethal capabilities to defeat remaining enemy forces in detail and reduce the will of those forces and the local population to resist. Divisions consolidate gains through decisive action focused on the defeat of bypassed forces and security between the close area and division rear boundary to maintain freedom of movement and the tempo of the operations.

Table 3-2. Consolidate gains by echelon

Echelon	Tasks to Consolidate Gains
Strategic-operational level (joint force land component commander—corps)	Establishing the security conditions necessary to achieve the desired political end state.
Operational-tactical level (field army—corps)	Exploiting tactical success to ensure the enemy cannot mount protracted resistance by other means.
Tactical level (corps—division)	Maintaining tempo and ensuring the enemy enjoys no respite; defeating the enemy in detail.

3-24. Consolidate gains is integral to the conclusion of all military operations, and it requires deliberate planning, preparation and resources to ensure sustainable success. This planning should ensure U.S. forces operate in a way that actively facilitates achievement of the desired post-hostilities end state and transition to legitimate authorities. Planners should anticipate task organization changes as conditions on the ground change over time, based on mission and operational variables. For example, additional engineer, military police, civil affairs, psychological operations, and sustainment capabilities are typically required to support the security and stability of large areas as they stabilize over time. In some instances, Army forces will be in charge of integrating and synchronizing these activities, in others Army forces will be in support. However, by backwards planning from the end state, and prioritizing the transition to legitimate authority, rather than simply planning from deployment of forces to the quick and efficient defeat of the enemy, commanders facilitate long-term success, limit mission creep, and minimize post-conflict problems.

ACTIVITIES TO CONSOLIDATE GAINS

3-25. To consolidate gains, Army forces take specific actions. These actions are described in paragraphs 3-26 through 3-36.

3-26. Army forces conduct a combination of offensive, defensive, and stability operations appropriate for their areas of operations. During combat, units consolidate gains in their areas of operations once large-scale ground combat has concluded in their area. Their initial focus is the defeat of all remaining enemy forces in detail and controlling all that could constitute a means for further resistance. This may require offensive action to defeat bypassed enemy units and secure enemy personnel, bases, equipment, and ammunition. It also requires an accurate understanding of enemy orders of battle and the capabilities that must be accounted for. As units establish area security, the balance of tasks should shift more heavily towards stability tasks focused on the control of populations and key nodes.

3-27. Area security is necessary to consolidate gains. Forces perform security tasks to protect friendly forces, routes, and critical infrastructure. Forces secure and control populations and enable freedom of friendly action within their area of operations.

3-28. Forces first perform minimum-essential stability operations tasks, then they establish a safe and secure environment to provide essential governmental services, emergency infrastructure reconstruction, and humanitarian relief. Maneuver forces may require significant augmentation to their task organization to perform stability tasks effectively.

3-29. Commanders ensure sufficient combat power is positioned within their area of operations to prevent counterattacks or infiltration of forces that could disrupt ongoing efforts to consolidate gains. These actions are especially important during operations intended to re-establish the international border for a friendly state, and they may be heavily weighted towards the defense.

3-30. Forces employ civil-military operations within their capabilities until they are augmented with civil affairs capabilities. The focus is to ensure minimal interference with friendly operations.

3-31. Forces employ information-related capabilities to influence the behavior of enemy forces. They also employ information-related capabilities with the local population in ways beneficial to achieve the desired end state.

3-32. Army forces consolidate gains through decisive action, initially weighted towards offensive operations against bypassed enemy forces and remnants of defeated enemy forces. They consolidate gains to ensure the area security essential to units operating in support areas to maintain offensive tempo. Doing so requires planning for additional forces so that commanders are not forced to shift combat power away from the close and deep areas. Maintaining offensive tempo requires additional combat power to conduct detention operations, relocate displaced civilians, establish law and order, provide humanitarian assistance, and secure key infrastructure. Ending enemy resistance and denying enemy forces the respite necessary to constitute a new form of resistance that prolongs the conflict are critical to success.

3-33. Commanders establish and sustain security during transitions between phases of operations to ensure there are minimal seams or gaps that allow enemy forces time to reorganize. Army forces perform continuous reconnaissance to gain or maintain contact with remaining enemy forces to enable their defeat and retain the initiative. Accounting for all enemy forces and their supporters helps determine the level of risk within their area of operations as well as the prioritization of tasks they assign subordinates. Commanders ensure that forces are properly task organized for the tasks they assign. Capabilities such as military information support operations, public affairs, and combat camera help in this effort.

3-34. Army forces are responsible for the provision of minimum-essential stability operations tasks. Generally these stability tasks include providing security, food, water, and medical treatment. However, Army forces may not perform all the essential tasks if another organization exists that can adequately perform those tasks. Army forces execute a greater number of stability tasks as requirements and capabilities evolve. The military retains the lead to establish civil security through the performance of security force assistance in all cases. The lead for the other all tasks eventually transfers to another military or civilian organization, although Army forces may retain a supporting role. (For more information on stability tasks, see ADP 3-07.)

3-35. Army forces must analyze the local capability and capacity to provide services as well as determine the ability of other U.S. government agencies, international agencies, NGOs, and contractors to provide support. The goal is to transition responsibility for humanitarian issues to entities other than Army forces as quickly as possible. This requires prior planning and coordination.

3-36. Consolidate gains may occur over a significant period and involve several changes in focus and emphasis as conditions change. An initial emphasis on defeating threat conventional forces will shift to more broadly based area security of populations and infrastructure. Eventually the emphasis and focus changes to meeting the needs of the population, influencing their perceptions, and allowing for a transition to a legitimate authority. Transitions are not generally abrupt, and units will manage different stability and security tasks concurrently until operations are complete. All activities should be prioritized towards securing and stabilizing the AO to meet the conditions necessary to achieve the desired conflict end state.

PRINCIPLES OF UNIFIED LAND OPERATIONS

3-37. A *principle* is a comprehensive and fundamental rule or an assumption of central importance that guides how an organization or function approaches and thinks about the conduct of operations (ADP 1-01). By integrating the six principles of unified land operations—mission command, develop the situation through action, combined arms, adherence to the law of war, establish and maintain security, and create multiple dilemmas for the enemy—Army commanders increase the probability of operational and strategic success. Success requires fully integrating U.S. military operations with the efforts of unified action partners. Success also requires commanders to exercise disciplined initiative to rapidly exploit opportunities that favorably develop the situation through action and create multiple dilemmas for the enemy.

Mission Command

3-38. The Army's command and control doctrine supports its operations doctrine. *Command and control* is the exercise of authority and direction by a properly designated commander over assigned and attached forces in the accomplishment of the mission (JP 1). Command and control is fundamental to the art and science of warfare. No single specialized military function, either by itself or combined with others, has a purpose without it. Through command and control, commanders provide purpose and direction to integrate all military activities towards a common goal—mission accomplishment.

3-39. *Mission command* is the Army's approach to command and control that empowers subordinate decision making and decentralized execution appropriate to the situation (ADP 6-0). Mission command enables the Army's operational concept of unified land operations and its emphasis on seizing, retaining, and exploiting the initiative. Mission command has seven fundamental principles:

- Competence.
- Trust.
- Shared understanding.
- Commander's intent.
- Mission orders.
- Disciplined initiative.
- Risk acceptance.

(See ADP 6-0 for a detailed discussion of the fundamental principles of mission command.)

3-40. The mission command approach to command and control is based on the Army's view that war is inherently chaotic and uncertain. No plan can account for every possibility, and most plans must change rapidly during execution to account for changes in the situation. No single person is ever sufficiently informed to make every important decision, nor can a single person keep up with the number of decisions that need to be made during combat. Enemy forces may behave differently than expected, a route may become impassable, or units could consume supplies at unexpected rates. Friction and unforeseeable combinations of variables impose uncertainty in all operations and demand an approach that does not attempt to impose perfect order, but rather accepts uncertainty and makes allowances for unpredictability.

3-41. Mission command helps commanders capitalize on subordinate ingenuity, innovation, and decision making to achieve the commander's intent when conditions change or current orders are no longer relevant. It requires subordinates who seek opportunities and commanders who accept risk for subordinates trying to meet their intent. Subordinate decision making and decentralized execution appropriate to the situation help manage uncertainty and enable necessary tempo at each echelon during operations.

3-42. Subordinates empowered to make decisions during operations unburden higher echelon commanders from issues that distract from necessary broader perspective and focus on critical issues. Mission command allows those commanders with the best situational understanding to make rapid decisions without waiting for higher echelon commanders to assess the situation and issue orders.

3-43. Decentralized execution is the delegation of decision making authority to subordinates, so they may make and implement decisions and adjust their assigned tasks in fluid and rapidly changing situations. Subordinate decisions should be ethically based, and within the framework of their higher echelon commander's intent. Decentralized execution is essential to seizing, retaining, and exploiting the initiative during operations in environments where conditions rapidly change and uncertainty is the norm. Rapidly changing situations and uncertainty are inherent in operations where commanders seek to establish a tempo and intensity that enemy forces cannot match.

3-44. Commanders determine the appropriate level of control, including delegating decisions and determining how much decentralized execution to employ. The level and application of control is constantly evolving and must be continuously assessed and adjusted to ensure the level of control is appropriate to the situation. Commanders should allow subordinates the greatest freedom of action commensurate with the level of acceptable risk in a particular situation.

3-45. Different operations and phases of operations may require tighter or more relaxed control over subordinate elements than others. Operations that require the close synchronization of multiple units, or the integration of effects in a limited amount of time, naturally require more detailed coordination, and may be controlled in a more centralized manner. Examples of this include combined arms breaches, air assaults, and wet gap crossings. Conversely, operations that do not require the close coordination of multiple units, such as a movement to contact or a pursuit, offer many opportunities to exercise initiative. These opportunities may be lost if too much emphasis is placed on detailed synchronization. Even in a highly controlled operation, subordinates must still exercise initiative to address unexpected problems and achieve their commander's intent when existing orders no longer make sense in the context of execution.

DEVELOP THE SITUATION THROUGH ACTION

3-46. During operations, commanders develop the situation through action. Developing the situation requires information. Commanders fight for information while in contact with enemy forces and gather information through close association with a population. Developing the situation through action to collect information is inherently part of displaying disciplined initiative. Commanders enhance situational awareness and understanding by assigning information collection tasks (reconnaissance, surveillance, security operations, and intelligence operations) to collect information requirements.

3-47. Often information can only be provided by close combat that forces enemy forces to reveal their capabilities, locations, and intent. When units encounter an enemy force or obstacle, then they must quickly determine the nature of the threat they face. Units share the enemy's dispositions, activities, and movements, along with an assessment, to their higher echelon headquarters and with the other units in their formation.

3-48. During planning, commanders identify information gaps, develop information requirements, and then assign collection tasks to subordinates. Information collection and analysis allows staffs to develop options for the commander who uses them to further seize opportunities and maintain initiative.

3-49. Commanders take enemy capabilities and reaction times into account when making decisions. They ensure that plans delegate decision-making authority to the lowest echelon possible to obtain faster and more suitable decisions. Subordinates use their initiative to further their higher echelon commander's intent.

3-50. During execution, commanders make decisions quickly, usually with incomplete information. Commanders who can make and implement decisions faster than enemy commanders, even to a small degree, gain an accruing advantage that becomes significant over time. Commanders should not delay a decision in hopes of finding a perfect solution to a military problem. By the time the slower commander decides and acts, the faster one has already altered the tactical situation, making the slower one's actions less effective or irrelevant. The faster commander maintains the initiative and dictates the tempo of operations.

3-51. To make timely decisions, commanders must understand the effects of their decisions in the context of their operational environment. They must understand enemy capabilities, the terrain and weather, and their impact on operations. Understanding an operational environment includes civil considerations—such as the population (with demographics and culture), the government, economics, NGOs, and history.

COMBINED ARMS

3-52. ***Combined arms* is the synchronized and simultaneous application of arms to achieve an effect greater than if each element was used separately or sequentially**. Combined arms integrates leadership, information, and each of the warfighting functions and joint capabilities through mission command. Used destructively, combined arms integrates different capabilities so that counteracting one makes the enemy vulnerable to another. Used constructively, combined arms uses all assets available to multiply the effectiveness and efficiency of Army capabilities used in stability or DSCA.

3-53. Combined arms uses the capabilities of all Army, joint, and multinational weapons systems—in the air, land, maritime, space, and cyberspace domains—in complementary and reinforcing ways. Complementary capabilities protect the weaknesses of one system or organization with the capabilities of a different one. During maneuver, the fires warfighting function complements the movement and maneuver warfighting function. Ground maneuver can make enemy forces vulnerable to joint weapon systems, while joint capabilities can enable maneuver. Electronic warfare capabilities can prevent enemy forces from communicating or relaying information about friendly maneuver. Information obtained from NGOs can facilitate effective distribution of supplies during humanitarian assistance and disaster relief operations.

3-54. Reinforcing capabilities combine similar systems or capabilities in the same warfighting function to increase the function's overall capabilities. In urban operations, for example, infantry, aviation, and armor units (movement and maneuver elements) working closely together reinforce the protection, maneuver, and direct fire capabilities of each unit type while creating cascading dilemmas for enemy forces. The infantry protects tanks from enemy infantry and antitank systems, while tanks provide protection and firepower for the infantry. Attack helicopters maneuver above buildings to protect ground formations, while other aircraft help sustain, extract, or air assault ground forces. Army artillery can be reinforced by close air support, air interdiction, air defense, and naval surface fire support, greatly increasing both the mass and range of fires available.

ADHERENCE TO THE LAW OF WAR

3-55. The *law of war* is that part of international law that regulates the conduct of armed hostilities (JP 1-04). The law of war's evolution was largely humanitarian and designed to reduce the evils of war. The main purposes of the law of war are to—
- Protect combatants, noncombatants, and civilians from unnecessary suffering.
- Provide certain fundamental protections for persons who fall into the hands of the enemy, particularly prisoners of war, civilians, and military wounded, sick, and shipwrecked.
- Facilitate the restoration of peace.
- Help military commanders in ensuring the disciplined and efficient use of military force.
- Preserve the professionalism and humanity of combatants.

3-56. Soldiers consider five important principles that govern the law of war when planning and executing operations: military necessity, humanity, distinction, proportionality, and honor. Three interdependent principles—military necessity, humanity, and honor—provide the foundation for other law of war principles—such as proportionality and distinction. Law of war principles work as interdependent and reinforcing parts of a coherent system. Military necessity justifies certain actions necessary to defeat enemy forces as quickly and efficiently as possible. Humanity forbids actions that cause unnecessary suffering. Proportionality requires that even when actions may be justified by military necessity, such actions may not be unreasonable or excessive. Distinction underpins the parties' responsibility to distinguish between the armed forces and the civilian population. Lastly, honor supports the entire system and gives parties confidence in it.

3-57. *Rules of engagement* are directives issued by competent military authority that delineate the circumstances and limitations under which United States forces will initiate and/or continue combat engagement with other forces encountered (JP 1-04). Rules of engagement always recognize the inherent right of self-defense. These rules vary between operations and types of units in the same area of operations, and may change during an operation. Adherence to them ensures Soldiers act consistently with international law, national policy, and military regulations.

3-58. Soldiers deployed to a combat zone overseas follow rules of engagement established by the Secretary of Defense and adjusted for theater conditions by the JFC. Within the United States and its territories, Soldiers adhere to rules for the use of force. Rules for the use of force consist of directives issued to guide U.S. forces during various operations. These directives may take the form of execute orders, deployment orders, memoranda of agreement, or plans. (See JP 3-28 for discussion on rules for the use of force.) Rules of engagement are permissive measures intended to allow the maximum use of destructive combat power appropriate for a mission. Rules for the use of force are restrictive measures intended to allow only the minimum force necessary to accomplish a mission. The underlying principle is a continuum of force, a carefully graduated level of response determined by the behavior of possible threats.

3-59. Successful operations require Army forces to employ lethal and nonlethal capabilities in a disciplined manner. Threats challenge the morals and ethics of Soldiers. Often an enemy does not respect international laws or conventions and commits atrocities simply to provoke retaliation in kind. Any loss of discipline on the part of friendly forces is likely to be distorted and exploited into propaganda, and magnified through the media. It is crucial that all personnel operate within applicable U.S., international, and in some cases host-nation laws and regulations. Ensuring friendly forces remain within legal, moral, and ethical boundaries is a leadership duty. This challenge rests heavily on small-unit and company-grade leaders charged with maintaining good order and discipline within their respective units. The Soldier's Rules in AR 350-1 distill the essence of the law of war. (See table 3-3 for a list of the Soldier's Rules.)

Table 3-3. The Soldier's Rules

- Soldiers fight only enemy combatants.
- Soldiers do not harm enemies who surrender. They disarm them and turn them over to their superior.
- Soldiers do not kill or torture any personnel in their custody.
- Soldiers collect and care for the wounded, whether friend or foe.
- Soldiers do not attack medical personnel, facilities, or equipment.
- Soldiers destroy no more than the mission requires.
- Soldiers treat civilians humanely.
- Soldiers do not steal. Soldiers respect private property and possessions.
- Soldiers should do their best to prevent violations of the law of war.
- Soldiers report all violations of the law of war to their superior.

ESTABLISH AND MAINTAIN SECURITY

3-60. Army forces perform area security to ensure freedom of action and to deny enemy forces the ability to disrupt operations. Commanders combine reconnaissance tasks and offensive, defensive, and stability operations to protect friendly forces, populations, infrastructure, and activities critical to mission accomplishment. Army forces integrate with partner military, law enforcement, and civil capabilities to establish and maintain security. The Army's ability to establish control is critical to consolidating gains in the wake of successful military operations.

3-61. Security operations prevent surprise, reduce uncertainty, and provide early warning of enemy activities. Warning provides friendly forces with time and maneuver space with which to react and develop the situation on favorable terms. Security operations prevent enemies from discovering the friendly plan and protect the force from unforeseen enemy actions. Security elements focus on preventing enemy forces from gathering essential elements of friendly information. Security is a dynamic effort that anticipates and thwarts enemy collection efforts. When successful, security operations allow the force to maintain the initiative.

CREATE MULTIPLE DILEMMAS FOR THE ENEMY

3-62. Simultaneous operations across multiple domains—conducted in depth and supported by military deception—present enemy forces with multiple dilemmas. These operations degrade enemy freedom of action, reduce enemy flexibility and endurance, and upset enemy plans and coordination. Such operations place critical enemy functions at risk and deny enemy forces the ability to synchronize or generate combat power. The application of capabilities in a complementary and reinforcing fashion creates more problems than the enemy commander can hope to solve, which erodes both enemy effectiveness and the will to fight.

3-63. Deception is a critical supporting enabler for creating multiple dilemmas, achieving operational surprise and maintaining the initiative. Successful deception operations degrade the ability of threat commanders to decide and act on accurate information. Deception inhibits effective enemy action by increasing the time, space, and resources necessary to understand friendly courses of action. Well executed deception creates a cumulative effect on decision-making cycles, and can cause inaction, delay, misallocation of forces, and surprise as enemy forces react to multiple real and false dilemmas. Deception is a force multiplier when properly resourced and executed.

3-64. Forcible entry operations can create multiple dilemmas by creating threats that exceed an enemy force's capability to respond. The capability to quickly project power across operational distances presents enemy forces with difficult decisions about how to array their forces in time and space. Rapid tactical maneuver to operational depth to exploit a penetration or envelopment creates similar effects.

3-65. Creating multiple dilemmas requires the recognition of opportunities to exploit. Understanding enemy dispositions and capabilities, and the characteristics of the terrain and population, informs situational understanding and course of action development. Employing mutually supporting forces along different axes to strike from unexpected directions creates dilemmas, particularly when Army and joint capabilities converge effects against enemy forces in multiple domains simultaneously. Commanders seek every opportunity to make enemy forces operate in different directions against massed capabilities at the time and locations of their choosing.

TENETS OF UNIFIED LAND OPERATIONS

3-66. *Tenets of operations* are desirable attributes that should be built into all plans and operations and are directly related to the Army's operational concept (ADP 1-01). The tenets are interrelated and mutually supporting. Tenets of unified land operations describe the Army's approach to generating and applying combat power across the range of military operations during decisive action. An operation is a military action, consisting of two or more related tactical actions designed to accomplish a strategic objective in whole or in part. A tactical action is a battle or engagement employing lethal and nonlethal actions designed for a specific purpose relative to the enemy, the terrain, friendly forces, or other entities. Operations can include an attack to seize a piece of terrain or destroy an enemy unit, the defense of a population, and the training of other militaries to enable security forces as part of building partner capacity. In the homeland, Army forces apply the tenets of operations when supporting civil authorities to save lives, alleviate suffering, and protect property. Army operations are characterized by four tenets:

- Simultaneity.
- Depth.
- Synchronization.
- Flexibility.

SIMULTANEITY

3-67. **Simultaneity is the execution of related and mutually supporting tasks at the same time across multiple locations and domains**. Army forces operating simultaneously across the air, land, maritime, space, and cyberspace domains presents dilemmas to adversaries and enemies, while reassuring allies and influencing neutrals. The simultaneous application of joint and combined arms capabilities across the range of military operations overwhelms the enemy physically and psychologically. Simultaneity requires creating shared understanding and purpose through collaboration with all elements of the friendly force. Commanders synchronize the employment of capabilities while balancing tempo against sustainment capacity to produce simultaneous results.

DEPTH

3-68. **Depth is the extension of operations in time, space, or purpose to achieve definitive results**. Army forces engage enemy forces throughout their depth, preventing the effective employment of reserves and disrupting command and control, logistics, and other capabilities not in direct contact with friendly forces. Operations in depth can disrupt the enemy's decision cycle. They contribute to protection by destroying enemy capabilities before enemy forces can use them. Empowering subordinates to act with initiative decentralizes decision making and increases tempo to achieve greater depth during operations.

3-69. Cyberspace operations, space-based capabilities, and psychological operations provide opportunities to engage adversaries and enemies across the depth of their formations. Each have planning considerations with regard to timing, authorities, and effects relative to physical actions in the land domain that should be factored into friendly courses of action.

SYNCHRONIZATION

3-70. *Synchronization* is the arrangement of military actions in time, space, and purpose to produce maximum relative combat power at a decisive place and time (JP 2-0). Synchronization is not the same as simultaneity; it is the ability to execute multiple related and mutually supporting tasks in different locations at the same time. These actions produce greater effects than executing each in isolation. For example, synchronization of information collection, obstacles, direct fires, and indirect fires results in the destruction of enemy formations during a defense. When conducting an offensive operation, synchronizing forces along multiple lines of operations forces enemy forces to distribute their capabilities instead of massing them.

3-71. Information networks and commander's intent enable synchronization. Networks facilitate situational awareness and rapid communication. Subordinate and adjacent units use their understanding of the commander's intent to synchronize their actions with other units without direct control from higher echelon headquarters. Neither networks nor commander's intent guarantee synchronization, but when used together they provide a powerful tool for leaders to synchronize their efforts.

3-72. Commanders determine the degree of control necessary to synchronize their operations. They balance synchronization with agility and initiative, but they never surrender the initiative for the sake of synchronization. Excessive synchronization can lead to too much control, which limits the initiative of subordinates and undermines mission command.

FLEXIBILITY

3-73. **Flexibility is the employment of a versatile mix of capabilities, formations, and equipment for conducting operations**. Commanders must be able to adapt to conditions as they change and employ forces in a variety of ways. Flexibility facilitates collaborative planning and decentralized execution. Leaders learn from experience (their own and that of others) and apply new knowledge to each situation. Flexible plans help units adapt quickly to changing circumstances in operations.

3-74. Flexibility and innovation are essential elements of any successful operation, and they are products of creative and adaptive leaders. Army forces continuously adapt as operational environments change across the range of military operations. Flexibility is a critical ingredient of mission analysis, plans, and operations.

SUCCESSFUL EXECUTION OF UNIFIED LAND OPERATIONS

3-75. Conducting unified land operations requires—
- A clear commander's intent and concept of operations that establishes the role of each element and its contribution to accomplishing the mission.
- A flexible and redundant command and control system.
- A shared understanding of an operational environment and the purpose of the operation.
- Proactive and continuous information collection and intelligence analysis.
- In depth planning for, and when authorized, conduct of cyberspace operations.
- Aggressive security operations.
- Rapid task organization and re-task organization.
- Disciplined initiative within the commander's intent.
- The ability to move quickly, operate dispersed, and sustain maneuver over distance.
- Planned, responsive, and anticipatory sustainment.
- Combat power applied through combined arms.
- Well-trained, cohesive teams and bold, imaginative leaders.
- Accepting risk as opportunity while mitigating risk to the mission and force.
- The ability to coordinate operations with unified action partners.
- The ability to consolidate gains.

3-76. Commanders change tactics, modify their exercise of command and control, change task organization, and adjust the weight placed on each element of decisive action throughout an operation. This helps to seize, retain, and exploit the initiative. Commanders base decisions on their understanding of the situation, available resources, and the force's ability to conduct operations. Commanders assess the progress of ongoing operations, changes in the situation, and the force's combat effectiveness. Commanders also assess how well a current operation is shaping conditions for subsequent missions.

This page intentionally left blank.

Chapter 4

Operations Structure

Chapter 4 discusses the operational framework that enables commanders to visualize and describe operations. It begins with the operations structure as a whole. Then it discusses the operations process. The chapter concludes with a discussion of the Army operational framework.

CONSTRUCT FOR OPERATIONS STRUCTURE

4-1. The operations structure consists of the operations process, combat power, and the operational framework. This is the Army's common construct for unified land operations. It allows Army leaders to organize efforts rapidly, effectively, and in a manner commonly understood across the Army. The operations process provides a broadly defined approach to developing and executing operations. The warfighting functions provide a common organization for critical functions. The operational framework provides Army leaders with conceptual options for arraying forces and visualizing and describing operations.

OPERATIONS PROCESS

4-2. The operations process is a commander-led activity informed by mission command principles. It consists of the major command and control activities performed during operations: planning, preparing, executing, and continuously assessing an operation. These activities may be sequential or simultaneous. They are rarely discrete and often involve a great deal of overlap. Commanders use the operations process to drive the planning necessary to understand, visualize, and describe their unique operational environments; make and articulate decisions; and direct, lead, and assess military operations. (See ADP 5-0 for a detailed discussion of the operations process.)

4-3. *Planning* is the art and science of understanding a situation, envisioning a desired future, and laying out effective ways of bringing that future about (ADP 5-0). Planning consists of two separate but interrelated components: a conceptual component and a detailed component. Successful planning requires the integration of both components. Army leaders employ three methodologies for planning: the Army design methodology, the military decision-making process, and troop leading procedures (described in paragraphs 4-14 through 4-18). Commanders determine how much of each methodology to use based on the scope of the problem, their familiarity with the methodology, the echelon, and the time available.

4-4. *Preparation* consists of those activities performed by units and Soldiers to improve their ability to execute an operation (ADP 5-0). Preparation creates conditions that improve friendly forces' opportunities for success. It requires commander, staff, unit, and Soldier actions to ensure the force is trained, equipped, and ready to execute operations. Preparation activities help commanders, staffs, and Soldiers understand a situation and their roles in upcoming operations and set conditions for successful execution.

4-5. *Execution* is the act of putting a plan into action by applying combat power to accomplish the mission and adjusting operations based on changes in the situation (ADP 5-0). Commanders and staffs use situational understanding to assess progress and make execution and adjustment decisions. In execution, commanders and staffs focus their efforts on translating decisions into actions. They apply combat power to seize, retain, and exploit the initiative to gain and maintain a position of relative advantage. This is the essence of unified land operations.

4-6. Finally, *assessment* is determination of the progress toward accomplishing a task, creating a condition, or achieving an objective (JP 3-0). Assessment precedes and then occurs during the other activities of the operations process. Assessment involves deliberately comparing forecasted outcomes with actual events to

Chapter 4

determine the overall effectiveness of force employment. Assessment helps commanders determine progress toward achieving a desired end state, accomplishing objectives, and performing tasks.

ARMY DESIGN METHODOLOGY IN THE OPERATIONS PROCESS

4-7. The Army design methodology is useful as an aid to conceptual thinking about unfamiliar problems. To produce executable plans, commanders integrate the Army design methodology with the detailed planning typically associated with the military decision-making process. Commanders who use the Army design methodology may gain a greater understanding of an operational environment and its problems. Once they have an understanding of the environment, they can better visualize an appropriate operational approach. This greater understanding allows commanders to provide a clear commander's intent and concept of operations.

4-8. Army design methodology is iterative, collaborative, and continuous. As the operations process unfolds, the commander, staff, subordinates, and other partners continue collaboration to improve their shared understanding. An improved understanding may lead to modifications to the commander's operational approach or an entirely new approach altogether. (See ATP 5-0.1 for more information on Army design methodology.)

THE MILITARY DECISION-MAKING PROCESS

4-9. The military decision-making process is an iterative planning methodology. It integrates activities of the commander, staff, subordinate headquarters, and other partners. This integration enables them to understand the situation and mission; develop, analyze, and compare courses of action; decide on the course of action that best accomplishes the mission; and produce an order for execution. The military decision-making process applies to both conceptual and detailed approaches. It is most closely associated with detailed planning.

4-10. For unfamiliar problems, executable solutions typically require integrating the Army design methodology with the military decision-making process. The military decision-making process helps leaders apply thoroughness, clarity, sound judgment, logic, and professional knowledge, so they understand situations, develop options to solve problems, and reach decisions. This process helps commanders, staffs, and others to think critically and creatively while planning. (See ADP 5-0 for more information on the military decision-making process.)

TROOP LEADING PROCEDURES

4-11. Troop leading procedures is a dynamic process used by small-unit leaders to analyze a mission, develop a plan, and prepare for an operation. Heavily weighted in favor of familiar problems and short planning periods, troop leading procedures are typically employed by organizations without staffs at the company level and below. Leaders use troop leading procedures to solve tactical problems when working alone or with a small group. For example, a company commander may use the executive officer, first sergeant, fire support officer, supply sergeant, and communications sergeant to help during troop leading procedures. (See ADP 5-0 for more information on troop leading procedures.)

COMBAT POWER

4-12. To execute operations, commanders conceptualize capabilities in terms of combat power. Combat power has eight elements: leadership, information, command and control, movement and maneuver, intelligence, fires, sustainment, and protection. The Army collectively describes the last six elements as warfighting functions. Commanders apply combat power through warfighting functions using leadership and information. (See chapter 5 for a discussion of combat power.)

ARMY OPERATIONAL FRAMEWORK

4-13. An operational framework is a cognitive tool that commanders and staffs use to visualize and describe the application of combat power, in time, space, purpose, and resources, as they develop the concept of operations. An operational framework organizes an area of geographic and operational responsibility for the commander and provides a way to describe the employment of forces. The framework illustrates the relationship between close operations, operations in depth, and other operations in time and space across domains.

4-14. As a visualization tool, the operational framework bridges the gap between a unit's conceptual understanding of the environment and its need to generate detailed orders that direct operations. Staffs anticipate changes in the situation over time and plan to add, subtract, or update elements of the operational framework. This contributes to shared understanding and helps inform the continuous adjustment of graphic control measures and other tools to direct operations. Due to the dynamic characteristics of operations, operational frameworks are dependent on the mission variables and not static templates.

4-15. The operational framework has four components. First, commanders are assigned an area of operations for the conduct of operations. Second, commanders can designate deep, close, support, and consolidation areas to describe the physical arrangement of forces in time and space. Third, within these areas commanders conduct decisive, shaping, and sustaining operations to articulate an operation in terms of purpose. Finally, commanders designate the main and supporting efforts to designate the shifting prioritization of resources.

AREA OF OPERATIONS

4-16. An *area of operations* is an operational area defined by a commander for land and maritime forces that should be large enough to accomplish their missions and protect their forces (JP 3-0). For land operations, an area of operations includes subordinate areas of operations assigned by Army commanders to their subordinate echelons. In operations, commanders use control measures to assign responsibilities, coordinate fire and maneuver, and control combat operations. A *control measure* is a means of regulating forces or warfighting functions (ADP 6-0). One of the most important control measures is the assigned area of operations. The Army commander or joint force land component commander is the supported commander within an area of operations designated by the JFC for land operations. Within their areas of operations, commanders integrate and synchronize combat power. To facilitate this integration and synchronization, commanders designate targeting priorities, effects, and timing within their areas of operations. Responsibilities within an area of operations include—

- Terrain management.
- Information collection, integration, and synchronization.
- Civil affairs operations.
- Movement control.
- Clearance of fires.
- Security.
- Personnel recovery.
- Airspace control.
- Minimum-essential stability operations tasks.

4-17. Commanders consider a unit's area of influence when assigning it an area of operations. An *area of influence* is a geographical area wherein a commander is directly capable of influencing operations by maneuver or fire support systems normally under the commander's command or control (JP 3-0). Understanding an area of influence helps the commander and staff plan branches to the current operation in which the force uses capabilities outside the area of operations. An area of operations should not be substantially larger than the unit's area of influence. Ideally, the area of influence would encompass the entire area of operations. An area of operations that is too large for a unit to effectively control increases risk, allowing sanctuaries for enemy forces and limiting joint flexibility.

4-18. An *area of interest* is that area of concern to the commander, including the area of influence, areas adjacent thereto, and extending into enemy territory (JP 3-0). This area also includes areas occupied by enemy forces who could jeopardize the accomplishment of the mission. An area of interest for stability or DSCA may be much larger than that area associated with the offense and defense. The area of interest always encompasses aspects of the air, cyberspace, and space domains, since capabilities resident in all three enable and affect operations on land.

Chapter 4

4-19. Areas of operations may be contiguous or noncontiguous. When they are contiguous, a boundary separates them. When areas of operations are noncontiguous, subordinate commands do not share a boundary. The higher echelon headquarters retains responsibility for areas not assigned to subordinate units.

DEEP, CLOSE, SUPPORT, AND CONSOLIDATION AREAS

4-20. **The *deep area* is where the commander sets conditions for future success in close combat.** Operations in the deep area involve efforts to prevent uncommitted enemy forces from being committed in a coherent manner. A commander's deep area generally extends beyond subordinate unit boundaries out to the limits of the commander's designated area of operations. The purpose of operations in the deep area is often tied to setting conditions for future events in time and space. Operations in the deep area might disrupt the movement of operational reserves or prevent enemy forces from employing long-range fires. In an operational environment where the enemy recruits insurgents from a population, deep operations might focus on interfering with the recruiting process, disrupting the training of recruits, or eliminating the underlying factors that enable the enemy to recruit. Planning for operations in the deep area includes considerations for information collection, airspace control, joint fires, obstacle emplacement, maneuver (air and ground), special operations, and information operations.

4-21. The higher echelon headquarters is responsible for deep areas within its area of operations. In some instances, a deep area may focus along a single line of operation. In other instances, a deep area may focus along multiple lines of operations. The mission variables of METT-TC inform the methods leaders use to direct operations in a deep area.

4-22. **The *close area* is the portion of the commander's area of operations where the majority of subordinate maneuver forces conduct close combat.** Operations in the close area are within a subordinate commander's area of operations. Commanders plan to conduct decisive operations using maneuver and fires in the close area, and they position most of the maneuver force in it. In the close area, depending on the echelon, one unit may conduct the decisive operation while others conduct shaping operations to fix a specific enemy formation or defeat remnants of by-passed or defeated enemy forces. Planning for operations in the close area includes fire control measures, movement control measures, maneuver, and obstacle emplacement. Operations in the close area are inherently lethal because they often involve direct fire engagements with enemy forces seeking to mass direct, indirect, and aerial fires against friendly forces.

4-23. A *support area* **is the portion of the commander's area of operations that is designated to facilitate the positioning, employment, and protection of base sustainment assets required to sustain, enable, and control operations**. Commanders assign a support area as a subordinate area of operations to support functions. It is where most of the echelon's sustaining operations occur. Within a division or corps support area, a designated unit such as a brigade combat team (BCT) or maneuver enhancement brigade provides area security, terrain management, movement control, mobility support, clearance of fires, and tactical combat forces for security. This allows sustainment units to focus on their primary function. Corps and divisions may have one or multiple support areas, located as required to best support the force. These areas may be non-contiguous to the other areas, in the close area, or in the consolidation area.

4-24. **The *consolidation area* is the portion of the land commander's area of operations that may be designated to facilitate freedom of action, consolidate gains through decisive action, and set conditions to transition the area of operations to follow on forces or other legitimate authorities**. Commanders establish a consolidation area, particularly in the offense as the friendly force gains territory, to exploit tactical success while enabling freedom of action for forces operating in the other areas. A consolidation area has all the characteristics of a close area, with the purpose to consolidate gains through decisive action once large-scale ground combat has largely ended in that particular area of operations.

4-25. The consolidation area does not necessarily need to surround—nor contain—the support area base clusters. It requires a purposefully task-organized, combined arms unit to perform area security and stability tasks and employ and clear fires. This unencumbers units conducting close operations and enables the higher echelon headquarters to focus on close operations, deep operations, and future planning. Corps and divisions may designate multiple consolidation areas. Units designated to a consolidation area conduct decisive action to defeat remnants of defeated or by-passed forces and stabilize the area for transition to legitimate authority. The forces necessary to consolidate gains represent a separate and distinct requirement beyond the BCTs and divisions required to conduct close and deep operations. To consolidate gains properly, the theater army plans and requests the additional required forces through the force-tailoring process.

DECISIVE, SHAPING, AND SUSTAINING OPERATIONS

4-26. Decisive, shaping, and sustaining operations lend themselves to a broad conceptual orientation. The ***decisive operation* is the operation that directly accomplishes the mission**. The decisive operation is the focal point around which commanders design an entire operation. The decisive operation is designed to determine the outcome of a major operation, battle, or engagement. Multiple subordinate units may be engaged in the same decisive operation across multiple domains. Decisive operations lead directly to the accomplishment of the commander's intent.

4-27. A ***shaping operation* is an operation at any echelon that creates and preserves conditions for success of the decisive operation through effects on the enemy, other actors, and the terrain.** Information operations, for example, may integrate engagement tasks into an operation to reduce tensions between Army units and different ethnic groups. In combat, synchronizing the effects of aircraft, artillery fires, and obscurants to delay or disrupt repositioning forces illustrates shaping operations. Shaping operations may occur throughout the area of operations and involve any combination of forces and capabilities across multiple domains. Shaping operations set conditions for the success of the decisive operation. Commanders may designate more than one shaping operation.

4-28. A ***sustaining operation* is an operation at any echelon that enables the decisive operation or shaping operations by generating and maintaining combat power.** Sustaining operations focus internally on friendly forces while decisive and shaping operations focus externally on the enemy or environment. Sustaining operations include personnel and logistics support, support area security, movement control, terrain management, and infrastructure development.

4-29. Sustaining operations are inseparable from decisive and shaping operations. Sustaining operations occur throughout the area of operations, not just within a support area. Failure to sustain results in mission failure. Sustaining operations determine endurance, tempo, and operational reach. They also determine how quickly Army forces reconstitute and how far Army forces can exploit success.

4-30. Throughout decisive, shaping, and sustaining operations, commanders and their staffs need to ensure that—
- Forces maintain positions of relative advantage.
- Operations are integrated with unified action partners.
- Continuity is maintained throughout operations.

Position of Relative Advantage

4-31. A ***position of relative advantage* is a location or the establishment of a favorable condition within the area of operations that provides the commander with temporary freedom of action to enhance combat power over an enemy or influence the enemy to accept risk and move to a position of disadvantage.** Positions of relative advantage may extend across multiple domains to provide opportunities for units to compel, persuade, or deter enemy decisions or actions. Commanders seek and create positions of advantage to exploit through action, and they continually assess friendly and enemy forces in relation to each other for opportunities to exploit. A key aspect in achieving a position of advantage is ***maneuver*, which is movement in conjunction with fires.**

4-32. Army forces must quickly both recognize and exploit positions of relative advantage because they are likely to be temporary when faced with capable, adaptive enemy forces attempting to gain a position of advantage over friendly forces. Commanders and staffs analyze relative capabilities across all domains that can influence an operation. Significant advantages in one domain can significantly offset disadvantages in another.

Integration

4-33. Army forces operate as a part of a larger joint and multinational effort with numerous unified action partners. Army leaders integrate Army capabilities within this larger effort. Commanders, enabled by their

Chapter 4

staffs, integrate numerous processes and activities within their formations and across the joint force. Integration involves efforts to operate with unified action partners and efforts to employ Army capabilities as part of the larger operational concept.

4-34. Army leaders use Army capabilities to complement those of their unified action partners. They also depend on partners' capabilities to supplement Army capabilities. Effective integration requires staffs to create a shared understanding and purpose through collaboration with unified action partners.

Continuity

4-35. Decision making during operations is continuous; it is not a discrete event. Commanders carefully balance priorities between current and future operations as part of controlling risk. They seek to accomplish the mission effectively while conserving resources for future operations. To maintain continuity of operations, commanders and staffs establish branches and sequels that facilitate future operations and reduce the risk inherent to transitions.

4-36. Commanders only make changes to plans when necessary. This presents subordinates with the fewest possible changes and the most time to spend on their own planning and execution. The fewer the changes, the less planning needed, and the greater the chance that the changes will be executed successfully.

4-37. When possible, commanders should ensure that changes do not preclude options for future operations. Normally this applies only to higher echelons with organic planning capabilities. Staffs develop options during planning, or commanders infer them based on their assessment of the current situation. Developing or inferring options depends on validating earlier assumptions and updating planning factors and running estimates. Future operations may be war-gamed using updated planning factors, estimates, and assumptions.

MAIN AND SUPPORTING EFFORTS

4-38. Commanders designate main and supporting efforts to establish clear priorities of support and resources among subordinate units. The ***main effort* is a designated subordinate unit whose mission at a given point in time is most critical to overall mission success.** It is usually weighted with the preponderance of combat power. Typically, commanders shift the main effort one or more times during execution. Designating a main effort temporarily prioritizes resource allocation. When commanders designate a unit as the main effort, it receives priority of support and resources to maximize combat power. Commanders establish clear priorities of support, and they shift resources and priorities to the main effort as circumstances and the commander's intent require. Commanders may designate a unit conducting a shaping operation as the main effort until the decisive operation commences. However, the unit with primary responsibility for the decisive operation then becomes the main effort upon the execution of the decisive operation.

4-39. A ***supporting effort* is a designated subordinate unit with a mission that supports the success of the main effort.** Commanders resource supporting efforts with the minimum assets necessary to accomplish the mission. Forces often realize success of the main effort through success of supporting efforts.

Chapter 5
Combat Power

This chapter discusses combat power. It first discusses the elements of combat power. The next section covers the six warfighting functions: command and control, movement and maneuver, intelligence, fires, sustainment, and protection. Lastly, the chapter discusses the means of organizing combat power.

THE ELEMENTS OF COMBAT POWER

5-1. ***Combat power* is the total means of destructive, constructive, and information capabilities that a military unit or formation can apply at a given time.** Operations executed through simultaneous offensive, defensive, stability, or DSCA operations require the continuous generation and application of combat power. To an Army commander, Army forces generate combat power by converting potential into effective action. Combat power includes all capabilities provided by unified action partners that are integrated and synchronized with the commander's objectives to achieve unity of effort in sustained operations.

5-2. To execute combined arms operations, commanders conceptualize capabilities in terms of combat power. Combat power has eight elements: leadership, information, command and control, movement and maneuver, intelligence, fires, sustainment, and protection. The elements facilitate Army forces accessing joint and multinational fires and assets. The Army collectively describes the last six elements as warfighting functions. Commanders apply combat power through the warfighting functions using leadership and information. (See figure 5-1.)

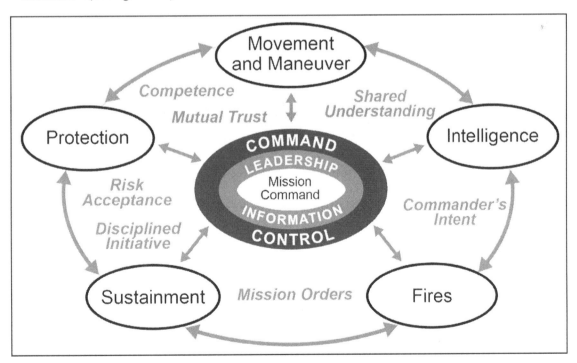

Figure 5-1. The elements of combat power

Chapter 5

5-3. Generating and maintaining combat power throughout an operation is essential. Factors that contribute to generating and maintaining combat power include reserves, force rotation, network viability, access to cyberspace and space enablers, and joint support. Commanders balance the ability to mass lethal and nonlethal effects with the need to deploy and sustain the units that produce those effects. They balance the ability to accomplish the mission with the ability to project and sustain the force.

5-4. Commanders apply leadership through mission command. Leadership is a multiplying and unifying element of combat power. The Army defines *leadership* as the activity of influencing people by providing purpose, direction, and motivation to accomplish the mission and improve the organization (ADP 6-22). An Army commander, by virtue of assumed role or assigned responsibility, inspires and influences people to accomplish organizational goals. (See ADP 6-22 for a detailed discussion of Army leadership.)

5-5. Information enables commanders at all levels to make informed decisions about the application of combat power and achieve definitive results. Knowledge management enables commanders to make informed, timely decisions under ambiguous and time-constrained conditions. Information management helps determine what among the vast amounts of information available is important. Information management uses procedures and information systems to facilitate collecting, processing, storing, displaying, disseminating, and protecting knowledge and information.

5-6. Commanders and their units must coordinate what they do, say, and portray. Fundamental to this coordination is the development of information themes and messages. An information theme is a unifying or dominant idea or image that expresses the purpose for military action. A *message* is a narrowly focused communication directed at a specific audience to support a specific theme (JP 3-61). Themes and messages are tied to objectives, lines of effort, and end state conditions. Information themes are overarching and apply to capabilities of public affairs, military information support operations, and audience engagements. Commanders employ themes and messages as part of planned activities designed to influence foreign audiences in support current or planned operations.

5-7. Every operation involves cyberspace electromagnetic activities. **Cyberspace electromagnetic activities is the process of planning, integrating, and synchronizing cyberspace and electronic warfare operations in support of unified land operations.** (This is also known as CEMA.) *Cyberspace operations* is the employment of cyberspace capabilities where the primary purpose is to achieve objectives in or through cyberspace (JP 3-0). *Electronic warfare* is military action involving the use of electromagnetic and directed energy to control the electromagnetic spectrum or to attack the enemy (JP 3-13.1).

5-8. Army cyberspace and electronic warfare operations are conducted to seize, retain, and exploit advantages in cyberspace and the electromagnetic spectrum. These operations support decisive action through the accomplishment of six core missions: offensive cyberspace operations, defensive cyberspace operations, DOD information network operations, electronic attack, electronic protection, and electronic warfare support. Commanders and staffs perform cyberspace electromagnetic activities to project power in cyberspace and the electromagnetic spectrum; secure and defend friendly force networks; and protect personnel, facilities, and equipment. Spectrum management operations are a critical enabler of integrated cyberspace operations and electronic warfare. (See FM 3-12 for a discussion of cyberspace operations and electronic warfare.)

THE SIX WARFIGHTING FUNCTIONS

5-9. A **warfighting function is a group of tasks and systems united by a common purpose that commanders use to accomplish missions and training objectives.** Warfighting functions are the physical means that tactical commanders use to execute operations and accomplish missions assigned by superior tactical- and operational-level commanders. The purpose of warfighting functions is to provide an intellectual organization for common critical capabilities available to commanders and staffs at all echelons and levels of warfare. Commanders integrate and synchronize these capabilities with other warfighting functions to accomplish objectives and missions.

5-10. All warfighting functions possess scalable capabilities to facilitate lethal and nonlethal effects. All the functions implement various systems such as personnel and networks to integrate forces and synchronize activities. Commanders should remember that cyber-related platforms that support integration and

Combat Power

synchronization must be protected and defended. Combined arms operations use the capabilities of each function, along with leadership and information, in complementary and reinforcing capabilities.

COMMAND AND CONTROL WARFIGHTING FUNCTION

5-11. **The *command and control warfighting function* is the related tasks and a system that enable commanders to synchronize and converge all elements of combat power.** The primary purpose of the command and control warfighting function is to assist commanders in integrating the other elements of combat power (leadership, information, movement and maneuver, intelligence, fires, sustainment, and protection) to achieve objectives and accomplish missions. The command and control warfighting function consists of the command and control warfighting function tasks and the command and control system. (See figure 5-2.)

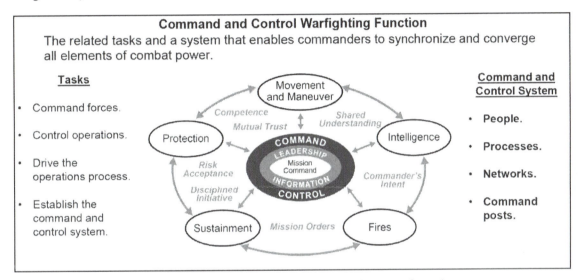

Figure 5-2. Command and control warfighting function

5-12. The command and control warfighting function tasks focus on integrating the activities of the other elements of combat power to accomplish missions. Commanders, assisted by their staffs, integrate numerous processes and activities within their headquarters and across the force through the command and control warfighting function. These tasks are—

- Command forces.
- Control operations.
- Drive the operations process.
- Establish the command and control system.

MOVEMENT AND MANEUVER WARFIGHTING FUNCTION

5-13. **The *movement and maneuver warfighting function* is the related tasks and systems that move and employ forces to achieve a position of relative advantage over the enemy and other threats.** Direct fire and close combat are inherent in maneuver. The movement and maneuver warfighting function includes tasks associated with force projection. Movement is necessary to disperse and displace the force as a whole or in part when maneuvering. Maneuver directly gains or exploits positions of relative advantage. Commanders use maneuver for massing effects to achieve surprise, shock, and momentum. Effective maneuver requires close coordination of fires and movement. Both tactical and operational maneuver require sustainment support. The movement and maneuver warfighting function includes these tasks:

- Move.
- Maneuver.
- Employ direct fires.
- Occupy an area.

- Conduct mobility and countermobility.
- Conduct reconnaissance and surveillance.
- Employ battlefield obscuration.

5-14. The movement and maneuver warfighting function does not include administrative movements of personnel and materiel. Those movements fall under the sustainment warfighting function. (See ADP 4-0 for a discussion of force projection.)

INTELLIGENCE WARFIGHTING FUNCTION

5-15. The *intelligence warfighting function* **is the related tasks and systems that facilitate understanding the enemy, terrain, weather, civil considerations, and other significant aspects of the operational environment.** Other significant aspects of an operational environment include threats, adversaries, and operational variables, which vary with the nature of operations. The intelligence warfighting function synchronizes information collection with primary tactical tasks of reconnaissance, surveillance, security, and intelligence operations. Intelligence is driven by commanders, and it involves analyzing information from all sources and conducting operations to develop the situation. The Army executes intelligence, surveillance, and reconnaissance through operations and intelligence processes, with an emphasis on intelligence analysis and information collection. The intelligence warfighting function includes these tasks:

- Provide intelligence support to force generation.
- Provide support to situational understanding.
- Conduct information collection.
- Provide intelligence support to targeting and information capabilities.

5-16. The intelligence warfighting function executes the tasks needed to prepare intelligence support to all echelons deployed within a theater of operation. There are three core tasks. First, the staff establishes and builds an intelligence architecture. Second, the staff builds the knowledge base needed to understand an operational environment through coordination and collaboration with regionally aligned forces using the theater military intelligence brigade. Building the knowledge to understand an operational environment includes connecting the intelligence architecture to theater information systems. Last, the staff supports engagement, develops context, and builds relationships through the successful conduct of intelligence operations; intelligence analysis; and intelligence processing, exploitation, and dissemination. (See ADP 2-0 for a discussion of the intelligence warfighting function and setting the theater.)

FIRES WARFIGHTING FUNCTION

5-17. The *fires warfighting function* **is the related tasks and systems that create and converge effects in all domains against the adversary or enemy to enable operations across the range of military operations (ADP 3-0).** These tasks and systems create lethal and nonlethal effects delivered from both Army and joint forces, as well as other unified action partners. The fires warfighting function does not wholly encompass, nor is it wholly encompassed by, any particular branch or function. Many of the capabilities that contribute to fires also contribute to other warfighting functions, often simultaneously. For example, an aviation unit may simultaneously execute missions that contribute to the movement and maneuver, fires, intelligence, sustainment, protection, and command and control warfighting functions.

5-18. Commanders must execute and integrate fires, in combination with the other elements of combat power, to create and converge effects and achieve the desired end state. Fires tasks are those necessary actions that must be conducted to create and converge effects in all domains to meet the commander's objectives. The tasks of the fires warfighting function are—

- Execute fires across the five domains and in the information environment, employing—
  - Surface-to-surface fires.
  - Air-to-surface fires.
  - Surface-to-air fires.
  - Cyberspace operations and electronic warfare.

- Space operations.
- Multinational fires.
- Special operations.
- Information operations.
- Integrate Army, multinational, and joint fires though—
  - Targeting.
  - Operations process.
  - Fire support planning.
  - Airspace planning and management.
  - Electromagnetic spectrum management.
  - Multinational integration.
  - Rehearsals.

The fires tasks are discussed further in ADP 3-19.

SUSTAINMENT WARFIGHTING FUNCTION

5-19. The sustainment warfighting function is one of the eight elements of combat power: leadership, information, command and control, movement and maneuver, intelligence, fires, sustainment, and protection. **The *sustainment warfighting function* is the related tasks and systems that provide support and services to ensure freedom of action, extended operational reach, and prolong endurance.** Sustainment determines the depth and duration of Army operations. Successful sustainment enables freedom of action by increasing the number of options available to the commander. Sustainment is essential for retaining and exploiting the initiative. The sustainment warfighting function consists of four elements:

- Logistics.
- Financial management.
- Personnel services.
- Health service support.

(See ADP 4-0 for additional information on the sustainment warfighting function.)

Logistics

5-20. *Logistics* is planning and executing the movement and support of forces. It includes those aspects of military operations that deal with: design and development; acquisition, storage, movement, distribution, maintenance, and disposition of materiel; acquisition or construction, maintenance, operation, and disposition of facilities; and acquisition or furnishing of services (ADP 4-0.) The explosive ordnance disposal tasks are discussed under the protection warfighting function. Army logistics elements are maintenance, transportation, supply, field services, distribution, operational contract support, and general engineering. (See FM 4-95 for additional information on logistics.)

Financial Management

5-21. Financial management leverages fiscal policy and economic power across the range of military operations. Financial management encompasses finance operations and resource management. (See FM 1-06 for additional details on financial management.)

Personnel Services

5-22. *Personnel services* are sustainment functions that man and fund the force, maintain Soldier and Family readiness, promote the moral and ethical values of the Nation, and enable the fighting qualities of the Army (ADP 4-0). Personnel services include planning, coordination, and sustaining personnel efforts at the operational and tactical levels. Personnel services include human resources support, legal support, religious support, and band support. (See ATP 1-19, FM 1-0, FM 1-04, and FM 1-05 for additional information on personnel services.)

Health Service Support

5-23. Army Health System support includes both health service support and force health protection which are critical capabilities embedded within formations across all warfighting functions. The force health protection mission falls under the protection warfighting function. (See ADP 3-37 for more information on the force health protection mission. See FM 4-02 for more information on health service support and the Army Health System.)

5-24. Health service support encompasses all support and services performed, provided, and arranged by the Army Medical Department to promote, improve, conserve, or restore the behavioral and physical well-being of Army personnel and as directed, unified action partners. Health service support includes casualty care, which encompasses a number of medical functions, including—

- Medical treatment (including organic and area medical support).
- Hospitalization.
- Dental care (including treatment aspects).
- Behavioral health and neuropsychiatric treatment.
- Clinical laboratory services.
- Treatment of chemical, biological, radiological, and nuclear patients.
- Medical evacuation (including medical regulating).
- Medical logistics (including blood management).

Protection Warfighting Function

5-25. The *protection warfighting function* **is the related tasks and systems that preserve the force so the commander can apply maximum combat power to accomplish the mission.** Commanders incorporate protection when they understand and visualize threats and hazards in an operational environment. This allows them to synchronize and integrate all protection capabilities to safeguard bases, secure routes, and protect forces. Preserving the force includes protecting personnel (combatants and noncombatants) and physical assets of the United States, unified action partners, and host nations. The protection warfighting function enables the commander to maintain the force's integrity and combat power. Protection determines the degree to which potential threats can disrupt operations to counter or mitigate those threats before they can act. However, protection is not a linear activity—planning, preparing, executing, and assessing protection is a continuous and enduring activity. Effective physical security measures, like any defensive measures, overlap and deploy in depth. Prioritization of protection capabilities are situationally dependent and resource-informed. Protection activities include developing and maintaining the protection prioritization list.

5-26. The protection warfighting function includes these tasks:
- Conduct survivability operations.
- Provide force health protection.
- Conduct chemical, biological, radiological, and nuclear operations.
- Provide explosive ordnance disposal support.
- Coordinate air and missile defense support.
- Conduct personnel recovery.
- Conduct detention operations.
- Conduct risk management.
- Implement physical security procedures.
- Apply antiterrorism measures.
- Conduct police operations.
- Conduct population and resource control.
- Conduct area security.
- Perform cyberspace security and defense.
- Conduct electromagnetic protection.
- Implement operations security.

(See ADP 3-37 for a discussion of the protection warfighting function.)

ORGANIZING COMBAT POWER

5-27. Commanders employ three means to organize combat power. These means include force tailoring, task-organizing, and mutual support.

FORCE TAILORING

5-28. **Force tailoring is the process of determining the right mix of forces and the sequence of their deployment in support of a joint force commander.** It involves selecting the right force structure for a joint operation from available units within a combatant command or from the Army force pool. Commanders then sequence forces into the area of operations as part of force projection. JFCs request and receive forces for each campaign phase, adjusting the quantity of Service component forces to match the weight of effort. Army Service component commanders tailor forces to meet land force requirements as determined by JFCs. Army Service component commanders also recommend forces and a deployment sequence to meet those requirements. Force tailoring is continuous. As new forces rotate into the area of operations, forces with excess capabilities return to the supporting combatant and Army Service component commands.

TASK ORGANIZATION

5-29. **Task-organizing is the act of designing a force, support staff, or sustainment package of specific size and composition to meet a unique task or mission.** Characteristics to examine when task-organizing the force include, but are not limited to, the mission, training, experience, unit capabilities, sustainability, the operational environment, and the enemy threat. Task-organizing includes allocating assets to subordinate commanders and establishing their command and support relationships. This occurs within tailored force packages as commanders organize subordinate units for specific missions and employ doctrinal command and support relationships. As task-organizing continues, commanders reorganize units for subsequent missions. The ability of Army forces to task-organize gives them extraordinary agility. It lets commanders configure their units to best use available resources. It also allows Army forces to match unit capabilities to tasks. The ability of sustainment forces to tailor and task-organize ensures commanders have freedom of action to change with mission requirements.

MUTUAL SUPPORT

5-30. Commanders consider mutual support when task-organizing forces, assigning areas of operations, and positioning units. Understanding mutual support and the time to accept risk during operations are fundamental to the art of tactics. In Army doctrine, mutual support is a planning consideration related to force disposition, not a command relationship. Mutual support has two aspects—supporting range and supporting distance. When friendly forces are static, supporting range equals supporting distance.

5-31. **Supporting range is the distance one unit may be geographically separated from a second unit yet remain within the maximum range of the second unit's weapons systems.** It depends on available weapons systems and is normally the maximum range of the supporting unit's indirect fire weapons. For small units (such as squads, sections, and platoons), it is the distance between two units that their direct fires can cover effectively. Visibility may limit the supporting range. If one unit cannot effectively or safely fire to support another, the first may not be in supporting range, even though its weapons have the required range.

5-32. **Supporting distance is the distance between two units that can be traveled in time for one to come to the aid of the other and prevent its defeat by an enemy or ensure it regains control of a civil situation.** These factors affect supporting distance:
- Terrain and mobility.
- Distance.
- Enemy capabilities.
- Friendly capabilities.
- Reaction time.

Chapter 5

5-33. The capabilities of supported and supporting units affect supporting distance. Units may be within supporting distance, but if the supported unit cannot communicate with the supporting unit, the supporting unit may not be able to affect the operation's outcome. In such cases, the units are not within supporting distance regardless of their proximity to each other. If the units share a common operational picture, relative proximity may be less important than both units' abilities to coordinate their maneuver and fires. To exploit the advantage of supporting distance, units synchronize maneuver and fires more effectively than enemy forces do. Otherwise, enemy forces may be able to defeat both units in detail.

5-34. Commanders consider the supporting distance in operations dominated by stability or DSCA tasks. Units maintain mutual support when one unit can draw on another unit's capabilities. An interdependent joint force may make proximity less significant than available capability. For example, Air Force assets may be able to move a preventive medicine detachment from an intermediate staging base to an operational area threatened by an epidemic.

5-35. Conventional and special operations forces may operate in proximity to each other to accomplish the JFC's mission. These two forces help and complement each other with mutual support so they can accomplish an objective that otherwise might not be attainable. Extended or large-scale operations involving both conventional and special operations forces require the integration and synchronization of conventional and special operations efforts. The JTF commander must consider the different capabilities and limitations of both conventional and special operations forces, particularly in the areas of command and control and sustainment. Exchanging liaison elements between conventional and special operations staffs further integrates efforts of all forces concerned. (For more information on coordinating conventional and special operations forces, see FM 6-05.)

5-36. Improved access to joint capabilities gives commanders additional means to achieve mutual support. Those capabilities can extend the operating distances between Army units. Army commanders can substitute joint capabilities for mutual support between subordinate forces. Using joint capabilities multiplies supporting distance many times over. Army forces can then extend operational reach over greater areas at a higher tempo. Joint capabilities are especially useful when subordinate units operate in noncontiguous areas of operations that place units beyond a supporting range or supporting distance. However, depending on joint capabilities outside an Army commander's direct control entails accepting risk when enemy forces can control multiple domains.

Appendix A
Command and Support Relationships

Command and support relationships provide the basis for unity of command and unity of effort in operations. All command and support relationships fall within the framework of joint doctrine. JP 1 discusses joint command relationships and authorities. Since Army support relationships differ from joint, commanders use Army support relationships when task-organizing Army forces.

COMMAND

A-1. Command is central to all military action, and unity of command is central to unity of effort. Inherent in command is the authority that a military commander lawfully exercises over subordinates, including the authority to assign missions and accountability for their successful completion. Although commanders may delegate authority to accomplish missions, they may not absolve themselves of the responsibility for the accomplishment of these missions. Authority is never absolute; the extent of authority is specified by the establishing authority, directives, and law. (See JP 1 for more information on command and authority.)

SUPPORT RELATIONSHIP

A-2. A support relationship is established by a common superior commander between subordinate commanders when one organization should aid, protect, complement, or sustain another force. The support relationship is used by the Secretary of Defense to establish and prioritize support between and among combatant commander, and it is used by JFCs to establish support relationships between and among subordinate commanders. (See JP 1 for more information on support relationships.)

CHAIN OF COMMAND

A-3. The President and Secretary of Defense exercise authority and control of the armed forces through two distinct branches of the chain of command. (See figure A-1 on page A-2.) One branch runs from the President, through the Secretary of Defense, to the combatant commanders for missions and forces assigned to combatant commands. The other branch runs from the President through the Secretary of Defense to the secretaries of the military departments. This branch is used for purposes other than operational direction of forces assigned to the combatant commands. Each military department operates under the authority, direction, and control of the secretary of that military department. These secretaries exercise authority through their respective Service chiefs over Service forces not assigned to combatant commanders. The Service chiefs, except as otherwise prescribed by law, perform their duties under the authority, direction, and control of the secretaries to whom they are directly responsible. (See JP 1 for more information on the chain of command.)

A-4. A typical operational chain of command extends from the combatant commander to a JTF commander, then to a functional component commander or a Service component commander. JTF and functional component commands, such as a land component, comprise forces that are normally subordinate to a Service component command but have been placed under the operational control (OPCON) of the JTF, and subsequently to a functional component commander. Conversely, the combatant commander may designate one of the Service component commanders as the JTF commander or as a functional component commander. In some cases, the combatant commander may not establish a JTF, retaining OPCON over subordinate functional commands and Service components directly.

Appendix A

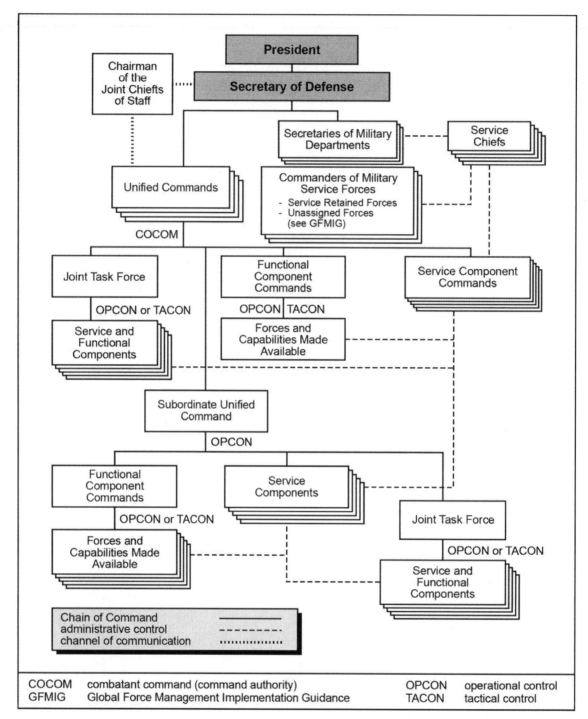

Figure A-1. Chain of command branches

A-5. Under joint doctrine, each joint force includes a Service component command that provides administrative and logistic support to Service forces under OPCON of that joint force. However, Army doctrine distinguishes between the Army component of a combatant command and Army components of subordinate joint forces. Under Army doctrine, Army Service component command (ASCC) refers to the Army component assigned to a combatant command. There is only one ASCC within a combatant command's area of responsibility. The Army components of all other joint forces are called ARFORs. An *ARFOR* is the Army component and senior Army headquarters of all Army forces assigned or attached to a combatant command, subordinate joint force command, joint functional command, or multinational command (FM 3-94). It consists of the senior Army headquarters and its commander (when not designated as the JFC) and all Army forces that the combatant commander subordinates to the JTF or places under the control of a multinational force commander. The ARFOR becomes the conduit for most Service-related issues and administrative support. The ASCC may function as an ARFOR headquarters when the combatant commander does not exercise command and control through subordinate JFCs.

A-6. The Secretary of the Army directs the flow of administrative control (ADCON). ADCON for Army units within a combatant command normally extends from the Secretary of the Army through the ASCC, through an ARFOR, and then to Army units assigned or attached to an Army headquarters within that joint command. However, ADCON is not tied to the operational chain of command. The Secretary of the Army may redirect some or all Service responsibilities outside the normal ASCC channels. In similar fashion, the ASCC may distribute some administrative responsibilities outside the ARFOR. Their primary considerations are the effectiveness of Army forces and the care of Soldiers.

COMBATANT COMMANDS

A-7. The Unified Command Plan establishes combatant commanders' missions and geographic responsibilities. Combatant commanders directly link operational military forces to the Secretary of Defense and the President. The Secretary of Defense deploys troops and exercises military power through the combatant commands. Six combatant commands have areas of responsibility. They are the geographic combatant commands. Each geographic combatant command has (or will have) an assigned ASCC. For doctrinal purposes, these commands become theater armies to distinguish them from the similar organizations assigned to functional component commands. The geographic combatant commands and their theater armies are

- U.S. Northern Command (U.S. Army, North-USARNORTH).
- U.S. Southern Command (U.S. Army, South-USARSO).
- U.S. Central Command (U.S. Army, Central-USARCENT).
- U.S. European Command (U.S. Army, Europe-USAREUR).
- U.S. Indo-Pacific Command (U.S. Army, Pacific-USARPAC).
- U.S. Africa Command (U.S. Army, Africa-USARAF).

In addition to these geographic combatant commands, U.S. Forces Korea is a subordinate unified command of U.S. Indo-Pacific Command. It also has a theater army (Eighth Army-EUSA).

A-8. There are four functional combatant commands. Each has global responsibilities. Like the geographic combatant commands, each has an ASCC assigned. These organizations are not theater armies; they are functional Service component commands. The functional combatant commands and their associated ASCCs are—

- U.S. Cyber Command (U.S. Army Cyber Command-ARCYBER).
- U.S. Strategic Command (U.S. Army Space and Missile Defense Command/Army Strategic Command-SMDC/ARSTRAT).
- U.S. Special Operations Command (U.S. Army Special Operations Command-USASOC).
- U.S. Transportation Command (Military Surface Deployment and Distribution Command-SDDC).

JOINT TASK FORCES AND SERVICE COMPONENTS

A-9. JTFs are the organizations most often used by a combatant commander for contingencies. Combatant commanders establish JTFs and designate the JFCs for these commands. Those commanders exercise OPCON of all U.S. forces through functional component commands, Service components, subordinate JTFs, or a combination of these. (See figure A-2 on page A-4.) The senior Army officer assigned to a JTF, other than the JFC and members of the JTF, becomes the ARFOR commander. The ARFOR commander answers to the Secretary of the Army through the ASCC for most ADCON responsibilities.

Appendix A

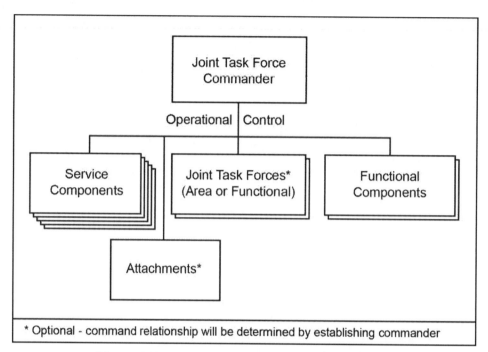

Figure A-2. Joint task force organization options

A-10. Depending on the JTF organization, the ARFOR commander may exercise OPCON of some or all Army forces assigned to the task force. For example, an Army corps headquarters may become a joint force land component within a large JTF. (See figure A-3, which shows an example of a JTF organized into functional components.) The corps commander exercises OPCON of Army divisions and tactical control (TACON) of Marine Corps forces within the land component. As the senior Army headquarters, the corps becomes the ARFOR for not only the Army divisions but also for all other Army units within the JTF, including those not under OPCON of the corps. This ensures that Service responsibilities are fulfilled while giving the JFC maximum flexibility for employing the joint force. Unless modified by the Secretary of the Army or the ASCC, Service responsibilities continue through the ARFOR to the respective Army commanders. Army forces in figure A-3 are shaded to show this relationship. The corps has OPCON of the Army divisions and TACON of the Marine division. The corps does not have OPCON over the other Army units but does, as the ARFOR, exercise ADCON over them. The corps also helps the ASCC in controlling Army support to other Services and to any multinational forces as directed.

A-11. When an Army headquarters becomes the joint force land component as part of a JTF, Army units subordinated to it are normally under OPCON. Marine Corps forces made available to a joint force land component command built around an Army headquarters are normally under TACON. The land component commander makes recommendations to the JFC on properly using attached, OPCON, or TACON assets; planning and coordinating land operations; and accomplishing such operational missions as assigned.

A-12. Navy and Coast Guard forces often operate under a joint force maritime component commander. This commander makes recommendations to the JFC on proper use of assets attached or under OPCON or TACON. Maritime component commanders also make recommendations concerning planning and coordinating maritime operations and accomplishing such missions.

A-13. A joint force air component commander is normally designated the supported commander for the air interdiction and counterair missions. The air component command is typically the headquarters with the most air assets. Like the other functional component commanders, the air component commander makes recommendations to the JFC on properly using assets attached or under OPCON or TACON. The air component commander makes recommendations for planning and coordinating air operations and accomplishing such missions. Additionally, the air component commander often has responsibility for airspace control authority and area air defense.

A-14. The JFC may organize special operations forces and conventional forces together as a joint special operations task force or a subordinate JTF. Other functional components and subordinate task forces—such as a joint logistic task force or joint psychological operations task force—are established as required.

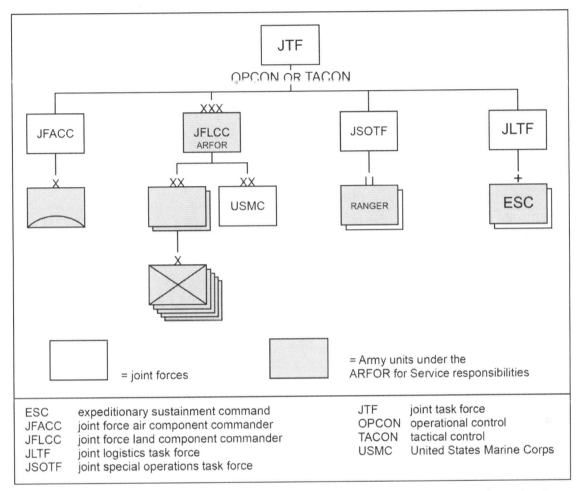

Figure A-3. Example of a joint task force showing an Army corps as joint force land component commander with ARFOR responsibilities

JOINT COMMAND RELATIONSHIPS

A-15. JP 1 specifies and details four types of joint command relationships:
- Combatant command (command authority) (COCOM).
- OPCON.
- TACON.
- Support.

Paragraphs A-16 through A-22 summarize important provisions of these relationships.

COMBATANT COMMAND (COMMAND AUTHORITY)

A-16. COCOM is the command authority over assigned forces vested only in commanders of combatant commands or as directed by the President or the Secretary of Defense in the Unified Command Plan and cannot be delegated or transferred. Title 10, United States Code, section 164 specifies it in law. Normally, the combatant commander exercises this authority through subordinate JFCs, Service component, and functional component commanders. COCOM includes the directive authority for logistic matters (or the

Appendix A

authority to delegate it to a subordinate JFC for common support capabilities required to accomplish the subordinate's mission).

OPERATIONAL CONTROL

A-17. OPCON is the authority to perform those functions of command over subordinate forces involving—
- Organizing and employing commands and forces.
- Assigning tasks.
- Designating objectives.
- Giving authoritative direction necessary to accomplish missions.

A-18. OPCON normally includes authority over all aspects of operations and joint training necessary to accomplish missions. It does not include directive authority for logistics or matters of administration, discipline, internal organization, or unit training. The combatant commander must specifically delegate these elements of COCOM. OPCON does include the authority to delineate functional responsibilities and operational areas of subordinate JFCs. In two instances, the Secretary of Defense may specify adjustments to accommodate authorities beyond OPCON in an establishing directive: when transferring forces between combatant commanders or when transferring members and organizations from the military departments to a combatant command. Adjustments will be coordinated with the participating combatant commanders. (See JP 1 for a detailed discussion of OPCON.)

TACTICAL CONTROL

A-19. TACON is inherent in OPCON. It may be delegated to and exercised by commanders at any echelon at or below the level of combatant command. TACON provides sufficient authority for controlling and directing the application of force or tactical use of maneuver support assets within the assigned mission or task. TACON does not provide organizational authority or authoritative direction for administrative and logistic support; the commander of the parent unit continues to exercise these authorities unless otherwise specified in the establishing directive. (See JP 1 for a detailed discussion of TACON.)

SUPPORT

A-20. Support is a command authority in joint doctrine. A supported and supporting relationship is established by a superior commander between subordinate commanders when one organization should aid, protect, complement, or sustain another force. Designating supporting relationships is important. It conveys priorities to commanders and staffs planning or executing joint operations. Designating a support relationship does not provide authority to organize and employ commands and forces, nor does it include authoritative direction for administrative and logistic support. Joint doctrine divides support into the categories listed in table A-1.

Command and Support Relationships

Table A-1. Joint support categories

Category	Definition
General support	That support which is given to the supported force as a whole and not to any particular subdivision thereof (JP 3-09.3).
Mutual support	That support which units render each other against an enemy, because of their assigned tasks, their position relative to each other and to the enemy, and their inherent capabilities (JP 3-31).
Direct support	A mission requiring a force to support another specific force and authorizing it to answer directly to the supported force's request for assistance (JP 3-09.3).
Close support	That action of the supporting force against targets or objectives that are sufficiently near the supported force as to require detailed integration or coordination of the supporting action (JP 3-31).

A-21. Support is, by design, somewhat vague but very flexible. Establishing authorities ensure both supported and supporting commanders understand the authority of supported commanders. Joint force commanders often establish supported and supporting relationships among components. For example, the maritime component commander is normally the supported commander for sea control operations; the air component commander is normally the supported commander for counterair operations. An Army headquarters designated as the land component may be the supporting force during some campaign phases and the supported force in other phases.

A-22. The JFC may establish a support relationship between functional and Service component commanders. Conducting operations across a large operational area often involves both the land and air component commanders. The JTF commander places the land component in general support of the air component until the latter achieves air superiority. Conversely, within the land area of operations, the land component commander becomes the supported commander and the air component commander provides close support. A joint support relationship is not used when an Army commander task-organizes Army forces in a supporting role. When task-organized to support another Army force, Army forces use one of four Army support relationships. (See paragraphs A-38 through A-39.)

JOINT ASSIGNMENT AND ATTACHMENT

A-23. Forces under the jurisdiction of the secretaries of the military departments (with some exceptions) are assigned to combatant commands or the commander, U.S. Element North America Aerospace Defense Command (known as USELEMNORAD). The exception exempts those forces necessary to carry out the functions of the military departments as noted in Title 10, United States Code, section 162. The assignment of forces to the combatant commands comes from the Secretary of Defense in the Forces for Unified Commands memorandum. According to this memorandum and the Unified Command Plan, unless otherwise directed by the President or the Secretary of Defense, all forces operating in the geographic area assigned to a combatant commander are assigned or attached to that combatant commander. A force assigned or attached to a combatant command may be transferred from that command to another combatant commander only when directed by the Secretary of Defense and approved by the President. The Secretary of Defense specifies the command relationship the gaining commander will exercise (and the losing commander will relinquish). Establishing authorities for subordinate unified commands and JTFs may direct the assignment or attachment of their forces to those subordinate commands and delegate the command relationship as appropriate. (See JP 1 for more information on joint assignment and attachment.)

A-24. When the Secretary of Defense assigns Army forces to a combatant command, the transfer is either permanent or the duration is unknown but very lengthy. The combatant commander exercises COCOM over assigned forces. When the Secretary of Defense attaches Army units, this indicates that the transfer of units is relatively temporary. Attached forces normally return to their parent combatant command at the end of the deployment. The combatant commander exercises OPCON of the attached force. In either case, the combatant commander normally exercises OPCON over Army forces through the ASCC until the combatant commander establishes a JTF or functional component. At that time, the combatant commander delegates OPCON to the JTF commander. When the JFC establishes any command relationship, the ASCC clearly specifies ADCON responsibilities for all affected Army commanders.

Appendix A

COORDINATING AUTHORITY

A-25. The *coordinating authority* is a commander or individual who has the authority to require consultation between the specific functions or activities involving forces of two or more Services, joint force components, or forces of the same Service or agencies, but does not have the authority to compel agreement (JP 1). In the event that essential agreement cannot be obtained, the matter shall be referred to the appointing authority. Coordinating authority is a consultation relationship, not an authority through which command may be exercised. Coordinating authority is more applicable to planning and similar activities than to operations. (See JP 1 for a detailed discussion of coordinating authority.) For example, a joint security commander exercises coordinating authority over area security operations within the joint security area. Commanders or leaders at any echelon at or below combatant command may be delegated coordinating authority. These individuals may be assigned responsibilities established through a memorandum of agreement between military and nonmilitary organizations.

DIRECT LIAISON AUTHORIZED

A-26. *Direct liaison authorized* is that authority granted by a commander (any level) to a subordinate to directly consult or coordinate an action with a command or agency within or outside of the granting command (JP 1). Direct liaison authorized is more applicable to planning than operations and always carries with it the requirement of keeping the commander granting direct liaison authorized informed. Direct liaison authorized is a coordination relationship, not an authority through which command may be exercised.

ADMINISTRATIVE CONTROL

A-27. *Administrative control* is direction or exercise of authority over subordinate or other organizations in respect to administration and support (JP 1). It is a Service authority, not a joint authority. It is exercised under the authority of and is delegated by the Secretary of the Army. ADCON is synonymous with the Army's Title 10 authorities and responsibilities.

A-28. The ASCC is always the senior Army headquarters assigned to a combatant command. Its commander exercises command authorities as assigned by the combatant commander and ADCON as delegated by the Secretary of the Army. ADCON is the Army's authority to administer and support Army forces even while in a combatant command area of responsibility. COCOM is the authority for command and control of the same Army forces. The Army is obligated to meet the combatant commander's requirements for the operational forces. Essentially, ADCON directs the Army's support of operational force requirements. Unless modified by the Secretary of the Army, administrative responsibilities normally flow from Department of the Army through the ASCC to those Army forces assigned or attached to that combatant command. ASCCs usually share ADCON for at least some administrative or support functions. Shared ADCON refers to the internal allocation of Title 10, United States Code, section 3013(b) responsibilities and functions. This is especially true for Reserve Component forces. Certain administrative functions, such as pay, stay with the Reserve Component headquarters, even after unit mobilization. Shared ADCON also applies to direct reporting units of the Army that typically perform single or unique functions. The direct reporting unit, rather than the ASCC, typically manages individual and unit training for these units. The Secretary of the Army directs shared ADCON.

ARMY COMMAND AND SUPPORT RELATIONSHIPS

A-29. Army command relationships are similar but not identical to joint command authorities and relationships. Differences stem from the way Army forces task-organize internally and the need for a system of support relationships between Army forces. Another important difference is the requirement for Army commanders to handle the administrative support requirements that meet the needs of Soldiers. These differences allow for flexible allocation of Army capabilities within various Army echelons. Army command and support relationships are the basis for building Army task organizations. A task organization is a temporary grouping of forces designed to accomplish a particular mission. Certain responsibilities are inherent in the Army's command and support relationships.

Army Command Relationships

A-30. Table A-2 on page A-10 lists the Army command relationships and possible authorities that can flow with those relationships. Command relationships define superior and subordinate relationships between units. Command relationships identify where the units or billets will reside in time and space. With a command relationship, separate command authority is transferred at discretion of the designating authority. When a chain of command is established, the command relationship unifies the effort and the specified authorities provide commanders the ability to employ subordinate forces with maximum flexibility. Army command relationships identify time and duration of assignment or attachment and authority identifies the degree of control provided to the gaining Army commander. The type of command relationship often relates to the expected longevity of the relationship between the headquarters involved, and it quickly identifies the degree of support that the gaining and losing Army commanders provide.

A-31. When referring to units, *organic* is assigned to and forming an essential part of a military organization as listed in its table of organization for the Army, Air Force, and Marine Corps, and are assigned to the operating forces for the Navy (JP 1). Joint command relationships do not include organic because a JFC is not responsible for the organizational structure of units. Joint units are formed and not recruited for manning, training and equipping. That is a service responsibility under Title 10 United States Code.

A-32. The Army establishes organic command relationships through organizational documents such as tables of organization and equipment and tables of distribution and allowances. If temporarily task-organized with another headquarters, organic units return to the control of their organic headquarters after accomplishing the mission. To illustrate, within a BCT, the entire brigade is organic. In contrast, within most modular support brigades, there is a base of organic battalions and companies and a variable mix of assigned and attached battalions and companies.

A-33. Army assigned units remain subordinate to the higher echelon headquarters for extended periods, typically years. Assignment is based on the needs of the Army, and it is formalized by orders rather than organizational documents. Although force tailoring or task-organizing may temporarily detach units, they eventually return to their either their headquarters of assignment or their organic headquarters. Attached units are temporarily subordinated to the gaining headquarters, and the period may be lengthy, often months or longer.

A-34. Attached units return to their parent headquarters (assigned or organic) when the reason for the attachment ends. The Army headquarters that receives another Army unit through assignment or attachment assumes responsibility for the ADCON requirements, and particularly sustainment, that normally extend down to that echelon, unless modified by directives or orders. For example, when an Army division commander attaches an engineer battalion to a BCT, the brigade commander assumes responsibility for the unit training, maintenance, resupply, and unit-level reporting for that battalion.

Appendix A

Table A-2. Command relationships

If relation-ship is—	Then inherent responsibilities—						
	Have command relationship with—	May be task-organized by—[1]	Unless modified, ADCON responsi-bility goes through—	Are assigned position or AO by—	Provide liaison to—	Establish and maintain communi-cations with—	Have priorities establish-ed by—	Authorities commander can impose on gaining unit further command or support relationship of—
Organic	All organic forces organized with the HQ	Organic HQ	Army HQ specified in organizing document	Organic HQ	N/A	N/A	Organic HQ	Attached; OPCON, TACON, GS, GSR, R, DS
Assigned	Gaining unit	Gaining HQ	Gaining Army HQ	OPCON chain of command	As required by OPCON	As required by OPCON	ASCC or Service-assigned HQ	As required by OPCON HQ
Attached	Gaining unit	Gaining unit	Gaining Army HQ	Gaining unit	As required by gaining unit	Unit to which attached	Gaining unit	Attached, OPCON, TACON, GS, GSR, R, DS
OPCON	Gaining unit	Parent unit and gaining unit, gaining unit may pass OPCON to lower echelon HQ[1]	Parent unit	Gaining unit	As required by gaining unit	As required by gaining unit and parent unit	Gaining unit	OPCON, TACON, GS, GSR, R, DS
TACON	Gaining unit	Parent unit	Parent unit	Gaining unit	As required by gaining unit	As required by gaining unit and parent unit	Gaining unit	TACON, GS, GSR, R, DS

Note: [1] In NATO, the gaining unit may not task-organize a multinational force. (See TACON.)

ADCON	administrative control	HQ	headquarters
AO	area of operations	N/A	not applicable
ASCC	Army Service component command	NATO	North Atlantic Treaty Organization
DS	direct support	OPCON	operational control
GS	general support	GSR	general support–reinforcing
TACON	tactical control	R	reinforcing

A-35. Army commanders normally place a unit OPCON or TACON to a gaining headquarters for a given mission, lasting perhaps a few days. OPCON lets the gaining commander task-organize and direct forces. TACON does not let the gaining commander task-organize the unit. Hence, TACON is the command relationship often used between Army, other Service, and multinational forces within a task organization, but rarely between Army forces. Neither OPCON nor TACON affects ADCON responsibilities. To modify the example used in paragraph A-34, if the Army division commander placed the engineer battalion OPCON to the BCT, the gaining brigade commander would not be responsible for the unit training, maintenance, resupply, and unit-level reporting of the engineers. Those responsibilities would remain with the parent maneuver enhancement brigade.

A-36. The ASCC and ARFOR monitor changes in joint organization carefully, and they may adjust ADCON responsibilities based on the situation. For example, if a JTF commander places an Army brigade under TACON of a Marine division, the ARFOR may switch some or all unit ADCON responsibilities to another Army headquarters, based on geography and ability to provide administration and support to that Army force.

Command and Support Relationships

ARMY SUPPORT RELATIONSHIPS

A-37. Table A-3 lists Army support relationships. Army support relationships are not a command authority, and they are more specific than the joint support relationships. Commanders establish support relationships when subordination of one unit to another is inappropriate. They assign a support relationship when—

- The support is more effective if a commander with the requisite technical and tactical expertise controls the supporting unit, rather than the supported commander.
- The echelon of the supporting unit is the same as or higher than that of the supported unit. For example, the supporting unit may be a brigade, and the supported unit may be a battalion. It would be inappropriate for the brigade to be subordinated to the battalion, hence the use of an Army support relationship.
- The supporting unit supports several units simultaneously. The requirement to set support priorities to allocate resources to supported units exists. Assigning support relationships is one aspect of command and control.

Table A-3. Army support relationships

If relation-ship is—	Then inherent responsibilities—							
	Have command relation-ship with—	May be task-organized by—	Receives sustain-ment from—	Are assigned position or an area of operations by—	Provide liaison to—	Establish and maintain communi-cations with—	Have priorities established by—	Authorities a command-er can impose on gaining unit further command or support relation-ship by—
Direct support[1]	Parent unit	Parent unit	Parent unit	Supported unit	Supported unit	Parent unit, supported unit	Supported unit	See note1
Reinforc-ing	Parent unit	Parent unit	Parent unit	Reinforced unit	Reinforced unit	Parent unit, reinforced unit	Reinforced unit, then parent unit	Not applicable
General support–reinforcing	Parent unit	Parent unit	Parent unit	Parent unit	Reinforced unit and as required by parent unit	Reinforced unit and as required by parent unit	Parent unit, then reinforced unit	Not applicable
General support	Parent unit	Parent unit	Parent unit	Parent unit	As required by parent unit	As required by parent unit	Parent unit	Not applicable

Note: [1] Commanders of units in direct support may further assign support relationships between their subordinate units and elements of the supported unit after coordination with the supported commander.

A-38. Army support relationships allow supporting commanders to employ their units' capabilities to achieve results required by supported commanders. Support relationships are graduated from an exclusive supported and supporting relationship between two units—as in direct support—to a broad level of support extended to all units under the control of the higher headquarters-as in general support. Support relationships do not alter ADCON. Commanders specify and change support relationships through task-organizing.

OTHER RELATIONSHIPS

A-39. Several other relationships established by higher echelon headquarters exist with units that are not in command or support relationships. (See table A-4 on page A-12.) These relationships are limited or

Appendix A

specialized to a greater degree than the command and support relationships. These limited relationships are not used when tailoring or task-organizing Army forces. Use of these specialized relationships helps clarify certain aspects of OPCON or ADCON.

Table A-4. Other relationships

Relationship	Operational use	Established by	Authority and limitations
TRO	TRO is an authority exercised by a combatant commander over assigned RC forces not on active duty. Through TRO, CCDRs shape RC training and readiness. Upon mobilization of the RC forces, TRO is no longer applicable.	The CCDR identified in the Forces for Unified Commands memorandum. The CCDR normally delegates TRO to the ASCC.	TRO allows the CCDR to provide guidance on operational requirements and training priorities, review readiness reports, and review mobilization plans for RC forces. TRO is not a command relationship. ARNG forces remain under the command and control of their respective State Adjutant Generals until mobilized for Federal service. USAR forces remain under the command and control of the USARC until mobilized.
Direct liaison authorized[1]	Allows planning and direct collaboration between two units assigned to different commands, often based on anticipated tailoring and task organization changes.	The parent unit headquarters. This is a coordination relationship, not an authority through which command may be exercised.	Limited to planning and coordination between units.
Aligned	Informal relationship between a theater army and other Army units identified for use in a specific geographic combatant command.	Theater army and parent ASCC.	Normally establishes information channels between the gaining theater army and Army units that are likely to be committed to that area of responsibility.

Note: [1] See also paragraph A-24.
ARNG   Army National Guard
ASCC   Army Service component command
CCDR   combatant commander
RC   Reserve Component
TRO   training and readiness oversight
USAR   U.S. Army Reserve
USARC   U.S. Army Reserve Command

A-40. *Training and readiness oversight* is the authority that combatant commanders may exercise over assigned Reserve Component forces when not on active duty or when on active duty for training (JP 1). Responsibilities for both training and readiness are inherent in ADCON, and they are exercised by unit commanders for their units. Army National Guard forces are organized by the Department of the Army under their respective states. These forces remain under command of the governor of that state until mobilized for Federal service. U.S. Army Reserve forces are assigned to U.S. Army Reserve Command. For Army National Guard units, combatant commanders normally exercise training and readiness oversight through their ASCC; for most, this is U.S. Army Forces Command. The ASCC coordinates with the appropriate State Adjutants General and Army National Guard divisions to refine mission-essential task lists for Army National Guard units. The ASCC coordinates mission-essential task lists for Army Reserve units with the U.S. Army Reserve

Command. When Reserve Component units align with an expeditionary force package during Army force generation, U.S. Army Forces Command establishes coordinating relationships as required between Regular Army and Reserve Component units. When mobilized, Reserve Component units are assigned or attached to their gaining headquarters. Most operating forces ADCON responsibilities, including unit training and readiness, shift to the gaining headquarters.

A-41. Army force packages for combatant commanders combine forces from many different parent organizations. The Army assigns or attaches Regular Army forces to various Army headquarters based on factors such as stationing, unit history, and habitual association of units in training. Different Army headquarters may share ADCON to optimize administration and support. For example, U.S. Army Forces Command may attach a BCT to a division headquarters located on a different installation. That division commander has training and readiness responsibilities for the BCT but does not control the training resources located at the BCT's installation. The senior Army commander on the BCT's installation manages training resources such as ranges and simulation centers. At the direction of the Secretary of the Army, the commanders share ADCON responsibilities. If the division headquarters deploys on an extended mission and the BCT remains, training and readiness responsibilities for the BCT shift to another commander. Headquarters, Department of the Army or another appropriate Army authority redistributes ADCON responsibilities for the BCT to a new headquarters. When the BCT deploys to a geographic combatant command, ADCON passes to the gaining theater army unless modified by the Secretary of the Army. (See FM 7-0 for a detailed discussion of training responsibilities.)

A-42. Alignment is an informal relationship between a theater army and other Army units identified for use in the area of responsibility of a geographic combatant command. Alignment helps focus unit exercises and other training on a particular region. This may lead to establishment of direct liaison authorized between the aligned unit and a different ASCC. Any modular Army force may find itself included in an expeditionary force package heading to a different combatant command. Therefore, Army commanders maintain a balance between regional focus and global capability.

ARFOR

A-43. The ARFOR is the Army component of any joint force. Army doctrine distinguishes, however, between the Army component of a combatant command and that of a joint force formed by the combatant commander. The Army component of the combatant command is an ASCC, and the Army component of the subordinate joint force is an ARFOR. (See JP 1 and JP 3-0 for more information on the ARFOR.)

THE ARFOR IN A SUBORDINATE JOINT FORCE

A-44. All JTFs that include Army forces have an ARFOR. The ARFOR consists of the Army commander, the commander's associated headquarters, and all Army forces attached to the JTF. The ARFOR provides administrative and logistics support to all Army forces and retains OPCON over Army units not subordinate to another component of the JTF. The senior Army officer assigned to the JTF, not in a joint duty assignment, becomes the ARFOR commander. Since the preferred joint approach for a JTF headquarters uses an existing Service headquarters, the JTF commander and headquarters retains all responsibilities associated with both command positions (ARFOR and JTF). This can overload the JTF headquarters unless the commander delegates authority for Service-specific matters to another commander. For example, when a corps becomes a JTF headquarters, the corps commander becomes the JTF commander. The corps retains ARFOR responsibilities through the ASCC back to the Army, unless the corps commander shifts Service responsibilities to another headquarters. The corps commander normally designates a subordinate Army commander and staff as the deputy ARFOR commander for performing those duties. (See JP 1 and JP 3-33 for more information on the ARFOR in a subordinate joint force.)

A-45. The typical JTF has a combination of Service and functional components. While the JTF will always have an ARFOR if it commands Army units, the operational roles of the ARFOR can vary. It is important to understand that the ARFOR exercises both OPCON and ADCON over Army forces in the JTF. However, not all Army forces are necessarily OPCON to the ARFOR. The ARFOR commander retains OPCON over Army forces attached to the joint force until the JFC places selected Army units under the command of another component in the JTF. The JFC may designate the senior Army commander and headquarters as the joint force land component command, in which case the Army commander exercises OPCON or TACON

Appendix A

over other Service forces, in addition to OPCON and ADCON over Army forces. In this case, dual command responsibilities as ARFOR and joint force land component are manageable, since the preponderance of forces are Army, and the missions assigned to other land forces are similar.

A-46. The combatant commander detaches Army forces from the theater army and attaches them to a JTF (or another joint force, such as a subunified command). This removes them from the OPCON of the theater army and places them under the OPCON of the gaining JFC. When command transfers to the gaining JTF, the ARFOR in the JTF exercises OPCON over Army forces attached to the JTF until the JFC directs otherwise. The JTF commander organizes the joint force by specifying command relationships (OPCON, TACON, or support) between attached forces. The ARFOR commander retains OPCON over those Army forces not subordinate to another component commander, such as a joint special operations component. The ARFOR commander is responsible for all aspects of planning and executing operations as directed by the JFC.

A-47. In addition to controlling Army forces, the ARFOR coordinates Army support to other Services (ASOS). ASOS includes provision of common-user logistics and executive agent support to the JTF as required by the JTF establishing authority. To make this coordination more manageable, the theater army normally retains command of logistics and medical support units that are not part of the brigades. These units provide area support not only to the Army forces but also to the joint force. The ARFOR headquarters manages support to other Services including, but not limited to—

- Missile defense.
- Fire support.
- Base defense.
- Transportation.
- Fuel distribution.
- General engineering.
- Intratheater medical evacuation.
- Veterinary services.
- Logistics management.
- Communications.
- Chemical, biological, radiological, and nuclear defense.
- Consequence management capability.
- Explosive ordnance disposal.

A-48. As required by the theater army, the ARFOR commander exercises ADCON over all Army forces in the JTF, including those subordinate to other components. Depending on the JTF organization, the ARFOR commander may exercise OPCON of some or all Army forces assigned to the task force, and the ARFOR commander remains responsible for ADCON of those forces. However, the exercise of OPCON is a delegation of joint command authority and not a function of ADCON.

A-49. The theater army commander will specify the ADCON responsibilities of the ARFOR, with the theater army normally retaining control of reception, staging, onward movement, and integration; logistics support of the deployed force; personnel support; and medical support. Administrative responsibilities normally retained by the ARFOR include internal administration and discipline, training within the joint operations area, and Service-specific reporting. The theater army normally retains OPCON of Army sustainment and medical commands operating in the joint operations area. The theater army commander establishes an Army support relationship between the ARFOR and these units.

ARFOR RESPONSIBILITIES

A-50. A division, corps, or field army headquarters serving as the ARFOR for a JTF includes the headquarters controlling multiple subordinate tactical formations and the Army forces placed under a joint or multinational headquarters. The Army commander is responsible to the JFC for these operational requirements. However, the JFC is not responsible for Service-specific matters involving administration and support of Army forces. The Army forces commander answers to the Secretary of the Army through the ASCC for Service-specific matters, whether it is a theater army or functional command (for example, the Surface Deployment and

Distribution Command of the United States Transportation Command). A theater army provides ADCON or Title 10 authorities and responsibilities for all Army units within the JTF, including those not under OPCON of the headquarters. In certain circumstances, such as geographic separation between the ADCON headquarters and the intermediate tactical headquarters, the theater army commander can delegate authority to execute specified administrative tasks to Army component commanders under OPCON of JFCs operating in joint operations areas within an area of operations.

A-51. The ARFOR within a joint operations area normally exercises OPCON over all Army maneuver, fires, and maneuver support forces (such as military police; air and missile defense; engineer; civil affairs; and chemical, biological, radiological, and nuclear), except for Army forces providing sustainment (including medical support). The ARFOR in a joint operations area identifies requirements, establishes priorities of support for Army forces, and coordinates with the theater army for providing sustainment. The ASCC may itself function as an ARFOR unless the combatant commander exercises command and control through subordinate JFCs. In this case, each subordinate JFC potentially has subordinate Army forces, while the ASCC exercises ADCON of all Army forces across the area of responsibility. The ASCC provides ASOS, common-user logistics, assignment eligibility and availability, and sustainment to interagency elements and Army, joint, and multinational forces in a joint operations area.

> **Army Executive Agent Responsibilities**
>
> Under the authority of the Secretary of Defense and Title 10, U.S. Code, the Army has been designated the executive agent by the Secretary of Defense or Deputy Secretary of Defense for foundational activities that are not necessarily landpower equities, but functions necessary to the entire joint force. These functions include, but are not limited to—
> - DOD detainee operations policy.
> - Armed Services blood program office.
> - Chemical and biological defense program.
> - Chemical demilitarization.
> - DOD combat feeding research and engineering program.
> - Defense Language Institute Foreign Language Center.
> - DOD level III corrections.
> - Explosives safety management.

A-52. This relationship relieves division or corps headquarters (as ARFOR within the joint operations area) of responsibility for directly exercising ADCON and sustaining tasks for Army forces and providing ASOS, common-user logistics, and Army executive agent responsibilities. The sustainment concept splits the responsibilities between the ARFOR in the joint operations area or the joint force command (division or corps) and the theater army. The theater army provides sustainment to all Army forces stationed in, transiting through, or operating within the area of operations. It also provides most ASOS, common-user logistics, and Army executive agent support to unified action partners within the area of operations. The theater army executes these sustainment responsibilities through its assigned theater sustainment command with expeditionary sustainment commands and tailored sustainment brigades provided from the Army pool of Service-retained rotational forces. The theater army provides medical services to support the force through its assigned medical command (deployment support) and forward deployed medical brigades. (See FM 4-02 for more information on medical support.)

A-53. For each of the combatant commands, the Secretary of Defense has assigned administrative and logistics support for subordinate joint elements to one of the four military services. The Department of the Army delegates its assigned Combatant Command Support Agent (CCSA) responsibilities to the respective theater army (or in the case of Korea, to the Eighth Army) for that geographic combatant command. U.S. Army South performs CCSA for USSOUTHCOM and SOCSOUTH. U.S. Army Europe performs CCSA for USEUCOM, U.S. Special Operations Command Europe, U.S. Army Africa performs CCSA for USAFRICOM, and U.S. Special Operations Command Africa. Eighth Army performs CCSA for U.S. Joint Forces Korea and U.S. Special Operations Command Korea. USASOC performs CCSA for Joint Special Operations Command. (See DODD 5100.03 for more information on combatant commands.)

REGULATORY AUTHORITIES

A-54. Regulations, policies, and other authoritative sources also direct and guide Army forces, Army commands, direct reporting units, ASCCs, and other Army elements. The Army identifies technical matters, such as network operations or contracting, and assigns responsibilities for them to an appropriate organization. These organizations use technical channels established by regulation, policy, or directive.

Appendix A

Commanders may also delegate authority for control of certain technical functions to staff officers or subordinate commanders. (See FM 6-0 for a detailed discussion of technical channels.)

A-55. The primary regulation governing the missions, functions, and command and staff relationships, including ADCON, of the subordinate elements of the Department of the Army is AR 10-87. This regulation prescribes the relationships and responsibilities among Army forces, Army commands, direct reporting units, and ASCCs. It includes channels for technical supervision, advice, and support for specific functions among various headquarters, agencies, and units. Other regulations and policies specify responsibilities in accordance with DOD directives and U.S. statutes.

Glossary

The glossary lists acronyms and terms with Army or joint definitions. Where Army and joint definitions differ, (Army) precedes the definition. The glossary lists terms for which ADP 3-0 is the proponent with an asterisk (*) before the term. For other terms, it lists the proponent publication in parentheses after the definition.

SECTION I – ACRONYMS AND ABBREVIATIONS

ADCON	administrative control
ADP	Army doctrine publication
AR	Army regulation
ASCC	Army Service component command
ASOS	Army support to other Services
ATP	Army techniques publication
BCT	brigade combat team
CCSA	Combatant Command Support Agent
COCOM	combatant command (command authority)
DA	Department of the Army
DOD	Department of Defense
DODD	Department of Defense directive
DSCA	defense support of civil authorities
FM	field manual
JFC	joint force commander
JP	joint publication
JTF	joint task force
METT-TC	mission, enemy, terrain and weather, troops and support available, time available, and civil considerations
NGO	nongovernmental organization
OPCON	operational control
TACON	tactical control
U.S.	United States

SECTION II – TERMS

adversary

A party acknowledged as potentially hostile to a friendly party and against which the use of force may be envisaged. (JP 3-0)

administrative control

Direction or exercise of authority over subordinate or other organizations in respect to administration and support. (JP 1)

Glossary

alliance
> The relationship that results from a formal agreement between two or more nations for broad, long-term objectives that further the common interests of the members. (JP 3-0)

area of influence
> A geographical area wherein a commander is directly capable of influencing operations by maneuver or fire support systems normally under the commander's command or control. (JP 3-0)

area of interest
> That area of concern to the commander, including the area of influence, areas adjacent thereto, and extending into enemy territory. (JP 3-0)

area of operations
> An operational area defined by a commander for land and maritime forces that should be large enough to accomplish their missions and protect their forces. (JP 3-0)

ARFOR
> The Army component and senior Army headquarters of all Army forces assigned or attached to a combatant command, subordinate joint force command, joint functional command, or multinational command. (FM 3-94)

Army design methodology
> A methodology for applying critical and creative thinking to understand, visualize, and describe problems and approaches to solving them. (ADP 5-0)

assessment
> Determination of the progress toward accomplishing a task, creating a condition, or achieving an objective. (JP 3-0)

base
> A locality from which operations are projected or supported. (JP 4-0)

base camp
> An evolving military facility that supports the military operations of a deployed unit and provides the necessary support and services for sustained operations. (ATP 3-37.10)

campaign
> A series of related operations aimed at achieving strategic and operational objectives within a given time and space. (JP 5-0)

center of gravity
> The source of power that provides moral or physical strength, freedom of action, or will to act. (JP 5-0)

*close area
> The portion of the commander's area of operations where the majority of subordinate maneuver forces conduct close combat.

*close combat
> Warfare carried out on land in a direct-fire fight, supported by direct and indirect fires and other assets.

close support
> That action of the supporting force against targets or objectives that are sufficiently near the supported force as to require detailed integration or coordination of the supporting action. (JP 3-31)

*combat power
> (Army) The total means of destructive, constructive, and information capabilities that a military unit or formation can apply at a given time.

*combined arms
> The synchronized and simultaneous application of arms to achieve an effect greater than if each element was used separately or sequentially.

command and control
> The exercise of authority and direction by a properly designated commander over assigned and attached forces in the accomplishment of the mission. (JP 1)

*command and control warfighting function
> The related tasks and a system that enable commanders to synchronize and converge all elements of combat power.

concept of operations
> A verbal or graphic statement that clearly and concisely expresses what the commander intends to accomplish and how it will be done using available resources. (JP 5-0)

*consolidate gains
> Activities to make enduring any temporary operational success and to set the conditions for a sustainable security environment, allowing for a transition of control to other legitimate authorities.

*consolidation area
> The portion of the land commander's area of operations that may be designated to facilitate freedom of action, consolidate gains through decisive action, and set conditions to transition the area of operations to follow on forces or other legitimate authorities.

control measure
> A means of regulating forces or warfighting functions. (ADP 6-0)

coordinating authority
> A commander or individual who has the authority to require consultation between the specific functions or activities involving forces of two or more Services, joint force components, or forces of the same Service or agencies, but does not have the authority to compel agreement. (JP 1)

culminating point
> The point at which a force no longer has the capability to continue its form of operations, offense or defense. (JP 5-0)

*cyberspace electromagnetic activities
> The process of planning, integrating, and synchronizing cyberspace and electronic warfare operations in support of unified land operations.

cyberspace operations
> The employment of cyberspace capabilities where the primary purpose is to achieve objectives in or through cyberspace. (JP 3-0)

*decisive action
> The continuous, simultaneous execution of offensive, defensive, and stability operations or defense support of civil authorities tasks.

*decisive operation
> The operation that directly accomplishes the mission.

decisive point
> A geographic place, specific key event, critical factor, or function that, when acted upon, allows commanders to gain a marked advantage over an enemy or contribute materially to achieving success. (JP 5-0)

*deep area
> Where the commander sets conditions for future success in close combat.

*defeat
> To render a force incapable of achieving its objectives.

*defeat mechanism
> A method through which friendly forces accomplish their mission against enemy opposition.

Glossary

defense support of civil authorities
Support provided by U.S. Federal military forces, DOD civilians, DOD contract personnel, DOD Component assets, and National Guard forces (when the Secretary of Defense, in coordination with the Governors of the affected States, elects and requests to use those forces in Title 32, United States Code, status) in response to requests for assistance from civil authorities for domestic emergencies, law enforcement support, and other domestic activities, or from qualifying entities for special events. (DODD 3025.18)

*defensive operation
An operation to defeat an enemy attack, gain time, economize forces, and develop conditions favorable for offensive or stability operations.

*depth
The extension of operations in time, space, or purpose to achieve definitive results.

destroy
A tactical mission task that physically renders an enemy force combat-ineffective until it is reconstituted. Alternatively, to destroy a combat system is to damage it so badly that it cannot perform any function or be restored to a usable condition without being entirely rebuilt. (FM 3-90-1)

direct liaison authorized
That authority granted by a commander (any level) to a subordinate to directly consult or coordinate an action with a command or agency within or outside of the granting command. (JP 1)

direct support
A mission requiring a force to support another specific force and authorizing it to answer directly to the supported force's request for assistance. (JP 3-09.3)

*disintegrate
To disrupt the enemy's command and control system, degrading its ability to conduct operations while leading to a rapid collapse of the enemy's capabilities or will to fight.

*dislocate
To employ forces to obtain significant positional advantage, rendering the enemy's dispositions less valuable, perhaps even irrelevant.

electronic warfare
Military action involving the use of electromagnetic and directed energy to control the electromagnetic spectrum or to attack the enemy. (JP 3-13.1)

end state
The set of required conditions that defines achievement of the commander's objectives. (JP 3-0)

*enemy
A party identified as hostile against which the use of force is authorized.

execution
The act of putting a plan into action by applying combat power to accomplish the mission and adjusting operations based on changes in the situation. (ADP 5-0)

*exterior lines
Lines on which a force operates when its operations converge on the enemy.

*fires warfighting function
The related tasks and systems that create and converge effects in all domains against the adversary or enemy to enable operations across the range of military operations.

*flexibility
The employment of a versatile mix of capabilities, formations, and equipment for conducting operations.

*force tailoring
> The process of determining the right mix of forces and the sequence of their deployment in support of a joint force commander.

foreign internal defense
> Participation by civilian and military agencies of a government in any of the action programs taken by another government or other designated organization to free and protect its society from subversion, lawlessness, insurgency, terrorism, and other threats to its security. (JP 3-22)

forward operating base
> An airfield used to support tactical operations without establishing full support facilities. (JP 3-09.3)

general support
> That support which is given to the supported force as a whole and not to any particular subdivision thereof. (JP 3-09.3)

hazard
> A condition with the potential to cause injury, illness, or death of personnel; damage to or loss of equipment or property; or mission degradation. (JP 3-33)

homeland defense
> The protection of United States sovereignty, territory, domestic population, and critical infrastructure against external threats and aggression or other threats as directed by the President. (JP 3-27)

*hybrid threat
> The diverse and dynamic combination of regular forces, irregular forces, terrorists, or criminal elements unified to achieve mutually benefitting effects

information environment
> The aggregate of individuals, organizations, and systems that collect, process, disseminate, or act on information. (JP 3-13)

*intelligence warfighting function
> The related tasks and systems that facilitate understanding the enemy, terrain, weather, civil considerations, and other significant aspects of the operational environment.

interagency coordination
> Within the context of Department of Defense involvement, the coordination that occurs between elements of Department of Defense, and participating United States Government departments and agencies for the purpose of achieving an objective. (JP 3-0)

*interior lines
> Lines on which a force operates when its operations diverge from a central point.

intermediate staging base
> A tailorable, temporary location used for staging forces, sustainment and/or extraction into and out of an operational area. (JP 3-35)

interorganizational cooperation
> The interaction that occurs among elements of the Department of Defense; participating United States Government departments and agencies; state, territorial, local, and tribal agencies; foreign military forces and government agencies; international organizations; nongovernmental organizations; and the private sector. (JP 3-08)

*isolate
> To separate a force from its sources of support in order to reduce its effectiveness and increase its vulnerability to defeat.

joint force
> A force composed of elements, assigned or attached, of two or more Military Departments operating under a single joint force commander. (JP 3-0)

Glossary

joint operations

Military actions conducted by joint forces and those Service forces employed in specified command relationships with each other, which of themselves, do not establish joint forces. (JP 3-0)

***landpower**

The ability—by threat, force, or occupation—to gain, sustain, and exploit control over land, resources, and people.

***large-scale combat operations**

Extensive joint combat operations in terms of scope and size of forces committed, conducted as a campaign aimed at achieving operational and strategic objectives.

***large-scale ground combat operations**

Sustained combat operations involving multiple corps and divisions.

law of war

That part of international law that regulates the conduct of armed hostilities. (JP 1-04)

leadership

The activity of influencing people by providing purpose, direction, and motivation to accomplish the mission and improve the organization. (ADP 6-22)

***line of effort**

(Army) A line that links multiple tasks using the logic of purpose rather than geographical reference to focus efforts toward establishing a desired end state.

***line of operations**

(Army) A line that defines the directional orientation of a force in time and space in relation to the enemy and links the force with its base of operations and objectives.

lodgment

A designated area in a hostile or potentially hostile operational area that, when seized and held, makes the continuous landing of troops and materiel possible and provides maneuver space for subsequent operations. (JP 3-18)

logistics

Logistics is planning and executing the movement and support of forces. It includes those aspects of military operations that deal with: design and development; acquisition, storage, movement, distribution, maintenance, and disposition of materiel; acquisition or construction, maintenance, operation, and disposition of facilities; and acquisition or furnishing of services. (ADP 4-0)

***main effort**

A designated subordinate unit whose mission at a given point in time is most critical to overall mission success.

major operation

A series of tactical actions (battles, engagements, strikes) conducted by combat forces, coordinated in time and place, to achieve strategic or operational objectives in an operational area. (JP 3-0)

***maneuver**

(Army) Movement in conjunction with fires.

measure of effectiveness

An indicator used to measure a current system state, with change indicated by comparing multiple observations over time. (JP 5-0)

measure of performance

An indicator used to measure a friendly action that is tied to measuring task accomplishment. (JP 5-0)

message

A narrowly focused communication directed at a specific audience to support a specific theme. (JP 3-61)

mission command

(Army) The Army's approach to command and control that empowers subordinate decision making and decentralized execution appropriate to the situation. (ADP 6-0)

*movement and maneuver warfighting function

The related tasks and systems that move and employ forces to achieve a position of relative advantage over the enemy and other threats.

multinational operations

A collective term to describe military actions conducted by forces of two or more nations, usually undertaken within the structure of a coalition or alliance. (JP 3-16)

mutual support

That support which units render each other against an enemy, because of their assigned tasks, their position relative to each other and to the enemy, and their inherent capabilities. (JP 3-31)

nongovernmental organization

A private, self-governing, not-for-profit organization dedicated to alleviating human suffering; and/or promoting education, health care, economic development, environmental protection, human rights, and conflict resolution; and/or encouraging the establishment of democratic institutions and civil society. (JP 3-08)

*offensive operation

An operation to defeat or destroy enemy forces and gain control of terrain, resources, and population centers.

operation

A sequence of tactical actions with a common purpose or unifying theme. (JP 1)

operational approach

A broad description of the mission, operational concepts, tasks, and actions required to accomplish the mission. (JP 5-0)

operational art

The cognitive approach by commanders and staffs—supported by their skill, knowledge, experience, creativity, and judgment—to develop strategies, campaigns, and operations to organize and employ military forces by integrating ends, ways, and means. (JP 3-0)

operational concept

A fundamental statement that frames how Army forces, operating as part of a joint force, conduct operations. (ADP 1-01)

operational environment

A composite of the conditions, circumstances, and influences that affect the employment of capabilities and bear on the decisions of the commander. (JP 3-0)

*operational initiative

The setting of tempo and terms of action throughout an operation.

operational reach

The distance and duration across which a force can successfully employ military capabilities. (JP 3-0)

organic

Assigned to and forming an essential part of a military organization as listed in its table of organization for the Army, Air Force, and Marine Corps, and are assigned to the operating forces for the Navy. (JP 1)

personnel services

Sustainment functions that man and fund the force, maintain Soldier and Family readiness, promote the moral and ethical values of the Nation, and enable the fighting qualities of the Army. (ADP 4-0)

*phase

(Army) A planning and execution tool used to divide an operation in duration or activity.

planning

The art and science of understanding a situation, envisioning a desired future, and laying out effective ways of bringing that future about. (ADP 5-0)

*position of relative advantage

A location or the establishment of a favorable condition within the area of operations that provides the commander with temporary freedom of action to enhance combat power over an enemy or influence the enemy to accept risk and move to a position of disadvantage.

preparation

Those activities performed by units and Soldiers to improve their ability to execute an operation. (ADP 5-0)

principle

A comprehensive and fundamental rule or an assumption of central importance that guides how an organization or function approaches and thinks about the conduct of operations. (ADP 1-01)

*protection warfighting function

The related tasks and systems that preserve the force so the commander can apply maximum combat power to accomplish the mission.

rules of engagement

Directives issued by competent military authority that delineate the circumstances and limitations under which United States forces will initiate and/or continue combat engagement with other forces encountered. (JP 1-04)

security cooperation

All Department of Defense interactions with foreign security establishments to build security relationships that promote specific United States security interests, develop allied and partner nation military and security capabilities for self-defense and multinational operations, and provide United States forces with peacetime and contingency access to allied and partner nations. (JP 3-20)

security force assistance

The Department of Defense activities that support the development of the capacity and capability of foreign security forces and their supporting institutions. (JP 3-20)

*shaping operation

An operation at any echelon that creates and preserves conditions for success of the decisive operation through effects on the enemy, other actors, and the terrain.

*simultaneity

The execution of related and mutually supporting tasks at the same time across multiple locations and domains.

*stability mechanism

The primary method through which friendly forces affect civilians in order to attain conditions that support establishing a lasting, stable peace.

*stability operation

An operation conducted outside the United States in coordination with other instruments of national power to establish or maintain a secure environment and provide essential governmental services, emergency infrastructure reconstruction, and humanitarian relief.

*support area

The portion of the commander's area of operations that is designated to facilitate the positioning, employment, and protection of base sustainment assets required to sustain, enable, and control operations.

*supporting distance

The distance between two units that can be traveled in time for one to come to the aid of the other and prevent its defeat by an enemy or ensure it regains control of a civil situation.

*supporting effort

A designated subordinate unit with a mission that supports the success of the main effort.

*supporting range

The distance one unit may be geographically separated from a second unit yet remain within the maximum range of the second unit's weapons systems.

*sustaining operation

An operation at any echelon that enables the decisive operation or shaping operations by generating and maintaining combat power.

*sustainment warfighting function

The related tasks and systems that provide support and services to ensure freedom of action, extend operational reach, and prolong endurance.

synchronization

The arrangement of military actions in time, space, and purpose to produce maximum relative combat power at a decisive place and time. (JP 2-0)

*task-organizing

The act of designing a force, support staff, or sustainment package of specific size and composition to meet a unique task or mission.

*tempo

The relative speed and rhythm of military operations over time with respect to the enemy.

tenets of operations

Desirable attributes that should be built into all plans and operations and are directly related to the Army's operational concept. (ADP 1-01)

*threat

Any combination of actors, entities, or forces that have the capability and intent to harm United States forces, United States national interests, or the homeland.

training and readiness oversight

The authority that combatant commanders may exercise over assigned Reserve Component forces when not on active duty or when on active duty for training. (JP 1)

unified action

The synchronization, coordination, and/or integration of the activities of governmental and nongovernmental entities with military operations to achieve unity of effort. (JP 1)

*unified action partners

Those military forces, governmental and nongovernmental organizations, and elements of the private sector with whom Army forces plan, coordinate, synchronize, and integrate during the conduct of operations.

*unified land operations

The simultaneous execution of offense, defense, stability, and defense support of civil authorities across multiple domains to shape operational environments, prevent conflict, prevail in large-scale ground combat, and consolidate gains as part of unified action.

unity of effort

Coordination and cooperation toward common objectives, even if the participants are not necessarily part of the same command or organization, which is the product of successful unified action. (JP 1)

*warfighting function

A group of tasks and systems united by a common purpose that commanders use to accomplish missions and training objectives.

This page intentionally left blank.

References

All websites accessed on 24 June 2019.

REQUIRED PUBLICATIONS

These documents must be available to the intended users of this publication.

DOD Dictionary of Military and Associated Terms. June 2019.

ADP 1-02. *Terms and Military Symbols.* 14 August 2018.

RELATED PUBLICATIONS

These sources contain relevant supplemental information.

JOINT PUBLICATIONS

Most joint publications are available online: https://www.jcs.mil/doctrine. Most Department of Defense publications are available at the Department of Defense Issuances Web site: https://dtic.mil/whs/directives.

DODD 3025.18. *Defense Support of Civil Authorities (DSCA).* 29 December 2010.

DODD 5100.03. *Support of the Headquarters of Combatant and Subordinate Unified Commands.* 9 February 2011.

JP 1. *Doctrine for the Armed Forces of the United States.* 25 March 2013.

JP 1-04. *Legal Support to Military Operations.* 2 August 2016.

JP 2-0. *Joint Intelligence.* 22 October 2013.

JP 3-0. *Joint Operations.* 17 January 2017.

JP 3-08. *Interorganizational Cooperation.* 12 October 2016.

JP 3-09.3. *Close Air Support.* 25 November 2014.

JP 3-10. *Joint Security Operations in Theater.* 13 November 2014.

JP 3-12. *Cyberspace Operations.* 8 June 2018.

JP 3-13. *Information Operations.* 27 November 2012.

JP 3-13.1. *Electronic Warfare.* 8 February 2012.

JP 3-16. *Multinational Operations.* 1 March 2019.

JP 3-18. *Joint Forcible Entry Operations.* 11 May 2017.

JP 3-20. *Security Cooperation.* 23 May 2017.

JP 3-22. *Foreign Internal Defense.* 17 August 2018.

JP 3-27. *Homeland Defense.* 10 April 2018.

JP 3-28. *Defense Support of Civil Authorities.* 29 October 2018.

JP 3-31. *Joint Land Operations.* 24 February 2014.

JP 3-33. *Joint Task Force Headquarters.* 31 January 2018.

JP 3-35. *Deployment and Redeployment Operations.* 10 January 2018.

JP 3-61. *Public Affairs.* 17 November 2015.

JP 4-0. *Joint Logistics.* 4 February 2019.

JP 5-0. *Joint Planning.* 16 June 2017.

References

ARMY PUBLICATIONS

Most Army doctrinal publications are available online: https://armypubs.army.mil/.

ADP 1-01. *Doctrine Primer*. 31 July 2019.
ADP 2-0. *Intelligence*. 31 July 2019.
ADP 3-07. *Stability*. 31 July 2019.
ADP 3-19. *Fires*. 31 July 2019.
ADP 3-28. *Defense Support of Civil Authorities*. 31 July 2019.
ADP 3-37. *Protection*. 31 July 2019.
ADP 3-90. *Offense and Defense*. 31 July 2019.
ADP 4-0. *Sustainment*. 31 July 2019.
ADP 5-0. *The Operations Process*. 31 July 2019.
ADP 6-0. *Mission Command*. 31 July 2019.
ADP 6-22. *Army Leadership*. 31 July 2019.
ADP 7-0. *Training Units and Developing Leaders*. 31 July 2019.
AR 10-87. *Army Commands, Service Component Commands, and Direct Reporting Units*. 11 December 2017.
AR 350-1. *Army Training and Leader Development*. 10 December 2017.
ATP 1-19. *Army Music*. 13 February 2015.
ATP 3-37.10/MCRP 3-40D.13. *Base Camps*. 27 January 2017.
ATP 4-10/MCRP 4-11 H/NTTP 4-09.1/AFMAN 10-409-O. *Multi-Service Tactics, Techniques, and Procedures for Operational Contract Support*. 18 February 2016.
ATP 5-0.1. *Army Design Methodology*. 1 July 2015.
FM 1-0. *Human Resources Support*. 1 April 2014.
FM 1-04. *Legal Support to the Operational Army*. 18 March 2013.
FM 1-05. *Religious Support*. 21 January 2019.
FM 1-06. *Financial Management Operations*. 15 April 2014.
FM 3-0. *Operations*. 6 October 2017.
FM 3-12. *Cyberspace and Electronic Warfare Operations*. 11 April 2017.
FM 3-16. *The Army in Multinational Operations*. 8 April 2014.
FM 3-22. *Army Support to Security Cooperation*. 22 January 2013.
FM 3-57. *Civil Affairs Operations*. 17 April 2019.
FM 3-90-1. *Offense and Defense Volume 1*. 22 March 2013.
FM 3-94. *Theater Army, Corps, and Division Operations*. 21 April 2014.
FM 4-02. *Army Health System*. 26 August 2013.
FM 4-95. *Logistics Operations*. 1 April 2014.
FM 6-0. *Commander and Staff Organization and Operations*. 5 May 2014.
FM 6-05/MCRP 3-30.4/NTTP 3-05.19/AFTTP 3-2.73/USSOCOM Pub 3-33. *Multi-Service Tactics, Techniques, and Procedures for Conventional Forces and Special Operations Forces Integration, Interoperability, and Interdependence*. 4 April 2018.
FM 7-0. *Train to Win in a Complex World*. 5 October 2016.
FM 27-10. *The Law of Land Warfare*. 18 July 1956.

UNITED STATES LAW

Most acts and public laws are available at https://www.congress.gov/.

Title 10, United States Code. *Armed Forces*.

Title 32, United States Code. *National Guard*.

PRESCRIBED FORMS

This section contains no entries.

REFERENCED FORMS

Unless otherwise indicated, DA forms are available on the Army Publishing Directorate Web site: https://armypubs.army.mil/.

DA Form 2028. *Recommended Changes to Publications and Blank Forms*.

This page intentionally left blank.

Index

Entries are by paragraph number.

A

activities to consolidate gains, 3-25–3-36
adherence to the law of war, 3-55–3-59
administrative control, A-27–A-28
    defined, A-27
adversary, defined, 1-18
alliance, defined, 1-50
area of influence, defined, 4-17
area of interest, defined, 4-18
area of operations, 4-16–4-19
    defined, 4-16
ARFOR, A-43–A-53
    in a subordinate joint force, A-44–A-49
    responsibilities, A-50–A-53
Army command and support relationships, A-29–A-43
Army command relationships, A-30–A-36
Army design methodology, defined, 2-7
    in the operations process, 4-7–4-8
Army operational framework, 4-13–4-39
Army strategic roles, 1-30–1-35
Army support relationships, A-37–A-38
Army's operational concept, 3-1–3-76
assessment, defined, 4-6
authorities, regulatory, A-54–A-55

B

base, defined, 2-59
base camp, defined, 2-59
basing, 2-59–2-64

C

campaign, defined, 1-54
campaign quality, 1-59–1-62
center of gravity, 2-27–2-30
    defined, 2-27

chain of command, A-3–A-6
close area, defined, 4-22
close combat, 1-63–1-66
    defined, 1-63
coalition, 1-51
combat power, 4-12, 5-1–5-36
    defined, 5-1
    elements of, 5-1–5-8
    organizing, 5-27–5-36
combatant command (command authority), A-16
combatant commands, A-7–A-8
combined arms, 3-52–3-54
    defined, 3-52
command, A-1
    chain of, A-3–A-6
command and control, defined, 3-38
command and control warfighting function, 5-11–5-12
    defined, 5-11
command and support relationships, A-1–A-55
command and support relationships, Army, A-29–A-43
command relationships, Army, A-30–A-36
concept of operations, defined, 2-5
consolidate gains, 3-21–3-24
    activities to, 3-25–3-36
    defined, 1-35
consolidation area, 4-23–4-24
    defined, 4-24
construct for operations structure, 4-1
continuity, 4-35–4-37
control measure, defined, 4-16
cooperation with civilian organizations, 1-44–1-48
coordinating authority, defined, A-25
create multiple dilemmas for the enemy, 3-62–3-65

culminating point, defined, 2-52
culmination, 2-52–2-53
cyberspace electromagnetic activities, defined, 5-7
cyberspace operations, defined, 5-7

D

decisive action, 3-3–3-20
    and homeland defense, 3-16–3-17
    defined, 3-3
    transitioning in, 3-18–3-20
decisive operation, defined, 4-26
decisive point, defined, 2-31
decisive points, 2-31–2-33
deep area, 4-20–4-21
    defined, 4-20
defeat, defined, 2-10
    mechanisms, 2-10–2-16
defeat mechanism, defined, 2-11
defense support of civil authorities, defined, 3-15
defensive operation, defined, 3-13
depth, 3-68–3-69
    defined, 3-68
destroy, defined, 2-12
develop the situation through action, 3-46–3-51
dilemmas, create multiple for the enemy, 3-62–3-65
direct liaison authorized, defined, A-26
disintegrate, defined, 2-14
dislocate, defined, 2-13

E

effort, lines of, 2-34–2-35, 2-37–2-39
electronic warfare, defined, 5-7
elements, of combat power, 5-1–5-8
    of operational art, 2-21–2-67
end state, defined, 1-54

Index

Entries are by paragraph number.

end state and conditions, 2-24–2-26
enemy, create multiple dilemmas for the, 3-62–3-65
    defined, 1-17
environment, operational, 1-1–1-25
environments, shape operational, 1-31–1-32
establish and maintain security, 3-60–3-61
execution, defined, 4-5
expeditionary capability, 1-59–1-62
exterior lines, defined, 2-36

F—G

financial management, 5-21
fires warfighting function, 5-17–5-18
    defined, 5-17
flexibility, 3-73–3-74
    defined, 3-73
force tailoring, defined, 5-28
foreign internal defense, defined, 1-42
forward operating base, defined, 2-63

H

hazard, defined, 1-20
hazards, and threats, 1-16–1-25
health service support, 5-23–5-24
homeland defense, defined, 3-16
    and decisive action, 3-16–3-17
human endeavor, war as a, 1-26–1-29
hybrid threat, defined, 1-19

I

information environment, defined, 1-9
integration, 4-33–4-34
intelligence warfighting function, 5-15–5-16
    defined, 5-15
interagency coordination, defined, 1-39
interior lines, defined, 2-36
intermediate staging base, defined, 2-62
interorganizational cooperation, defined, 1-40
isolate, defined, 2-15

J—K

joint assignment and attachment, A-23–A-24
joint command relationships, A-15–A-28
joint force, defined, 1-54
joint operations, defined, 1-54
    principles of, 2-3
joint task forces and service components, A-9–A-14

L

land operations, 1-55–1-66
landpower, defined, 1-58
large-scale combat operations, defined, 1-28
large-scale ground combat, prevail in, 1-34
large-scale ground combat operations, defined, 1-7
law of war, adherence to the, 3-55–3-59
    defined, 3-55
leadership, defined, 5-4
line of effort, defined, 2-37
line of operations, defined, 2-36
lines of effort, 2-34–2-35, 2-37–2-39
lines of operations, 2-34–2-36, 2-39
lodgment, defined, 2-64
logistics, defined, 5-20

M

main effort, defined, 4-38
major operation, defined, 1-54
maneuver, defined, 4-31
measure of effectiveness, defined, 2-38
measure of performance, defined, 2-38
mechanisms, defeat, 2-10–2-16
    stability, 2-17–2-20
message, defined, 5-6
military decision-making process, 4-9–4-10
military operations, 1-1–1-75
mission command, 3-38–3-45
    defined, 3-39
mission variables, 1-13, 1-15
movement and maneuver warfighting function, 5-13–5-14
    defined, 5-13

multinational operations, 1-49–1-53
    defined, 1-49
mutual support, 5-30

N

nongovernmental organization, defined, 1-46

O

offensive operation, defined, 3-12
operation, defensive, 3-13
    defined, 1-55
    major, 1-54
    offensive, 3-12
    stability, 3-14
operational environment, 1-1–1-25
operational approach, defined, 2-8
operational art, 2-1–2-67
    defined, 2-1
    elements of, 2-21–2-67
operational concept, the Army's, 3-1–3-76
operational control, A-17–A-18
operational environment, defined, 1-1
operational initiative, defined, 1-67
operational reach, defined, 1-60
operational research, 2-54–2-58
operational variables, 1-13–1-14
operations, joint, 1-54
    land, 1-55–1-66
    lines of, 2-34–2-36, 2-39
    military, 1-1–1-75
    multinational, 1-49–1-53
operations process, 4-2–4-10
    Army design methodology in, 4-7–4-8
operations structure, 4-1–4-39
    construct for, 4-1
organic, defined, A-31
organizing combat power, 5-27–5-36
other relationships, A-39–A-42

P—Q

personnel services, defined, 5-22
phase, defined, 2-44
phasing and transitions, 2-44–2-51
planning, defined, 4-3
position of relative advantage, 4-31–4-32
    defined, 4-31

Entries are by paragraph number.

preparation, defined, 4-4
prevail in large-scale ground combat, 1-34
prevent conflict, 1-33
principle, defined, 3-37
principles, of joint operations, 2-3
  of unified land operations, 3-37–3-65
protection warfighting function, 5-25–5-26
  defined, 5-25

R

readiness through training, 1-71–1-75
regulatory authorities, A-54–A-55
relationships, command and support, A-1–A-55
responsibilities, ARFOR, A-50–A-53
risk, defined, 2-65–2-67
rules of engagement, defined, 3-57

S

security, establish and maintain, 3-60–3-61
security cooperation, defined, 1-41
security force assistance, defined, 1-42
seize, retain, and exploit the operational initiative, 1-67–1-70
service components, joint task forces and, A-9–A-14
shape operational environments, 1-31–1-32
shaping operation, defined, 4-27

simultaneity, 3-10–3-11
  defined, 3-67
six warfighting functions, 5-9–5-26
Soldier's Rules, 3-59
stability mechanism, defined, 2-19
stability mechanisms, 2-17–2-20
stability operation, defined, 3-14
stability tasks, 2-17
strategic roles, Army, 1-30–1-35
subordinate joint force, ARFOR in a, A-44–A-49
successful execution of unified land operations, 3-75–3-76
support, A-20–A-22
support area, defined, 4-23
support relationship, A-2
support relationships, Army, A-37–A-38
supporting distance, 5-32
supporting effort, defined, 4-39
supporting range, defined, 5-31
sustaining operation, 4-28–4-30
  defined, 4-28
sustainment warfighting function, defined, 5-19
synchronization, 3-70–3-72
  defined, 3-70

T

tactical control, A-19
task-organizing, defined, 5-29
tempo, 2-40–2-43
  defined, 2-40
tenets of operations, defined, 3-66
tenets of unified land operations, 3-66–3-74

threat, defined, 1-16
threats and hazards, 1-16–1-25
training and readiness oversight, defined, A-40
transitioning in decisive action, 3-18–3-20
transitions, and phasing, 2-44–2-51
troop leading procedures, 4-11

U

unified action, 1-36–1-54
  defined, 1-36
unified action partners, defined, 1-36
unified land operations, 3-1–3-2
  defined, 3-1
  principles of, 3-37–3-65
  successful execution of, 3-75–3-76
  tenets of, 3-66–3-74
unity of effort, defined, 1-36

V

variables, mission, 1-13–1-15
  operational, 1-13–1-14

W—X—Y—Z

war as a human endeavor, 1-26–1-29
warfighting function, defined, 5-9
warfighting functions, 5-9–5-26

Made in the USA
Columbia, SC
29 July 2024